Lecture Notes in Applied and Computational Mechanics

Volume 51

Series Editors

Prof. Dr.-Ing. Friedrich Pfeiffer
Prof. Dr.-Ing. Peter Wriggers

Lecture Notes in Applied and Computational Mechanics

Edited by F. Pfeiffer and P. Wriggers

Further volumes of this series found on our homepage: springer.com

Vol.51: Besdo, D., Heimann, B., Klüppel, M.,
Kröger, M., Wriggers, P., Nackenhorst, U.
Elastomere Friction
249 p. 2010 [978-3-642-10656-9]

Vol.50: Ganghoffer, J.-F., Pastrone, F. (Eds.)
Mechanics of Microstructured Solids 2
102 p. 2010 [978-3-642-05170-8]

Vol. 49: Hazra, S.B.
Large-Scale PDE-Constrained Optimization
in Applications
224 p. 2010 [978-3-642-01501-4]

Vol. 48: Su, Z.; Ye, L.
Identification of Damage Using Lamb Waves
346 p. 2009 [978-1-84882-783-7]

Vol. 47: Studer, C.
Numerics of Unilateral Contacts and Friction
191 p. 2009 [978-3-642-01099-6]

Vol. 46: Ganghoffer, J.-F., Pastrone, F. (Eds.)
Mechanics of Microstructured Solids
136 p. 2009 [978-3-642-00910-5]

Vol. 45: Shevchuk, I.V.
Convective Heat and Mass Transfer in Rotating Disk
Systems
300 p. 2009 [978-3-642-00717-0]

Vol. 44: Ibrahim R.A., Babitsky, V.I., Okuma, M. (Eds.)
Vibro-Impact Dynamics of Ocean Systems and Felated
Problems
280 p. 2009 [978-3-642-00628-9]

Vol.43: Ibrahim, R.A.
Vibro-Impact Dynamics
312 p. 2009 [978-3-642-00274-8]

Vol. 42: Hashiguchi, K.
Elastoplasticity Theory
432 p. 2009 [978-3-642-00272-4]

Vol. 41: Browand, F., Ross, J., McCallen, R. (Eds.)
Aerodynamics of Heavy Vehicles II: Trucks, Buses,
and Trains
486 p. 2009 [978-3-540-85069-4]

Vol. 40: Pfeiffer, F.
Mechanical System Dynamics
578 p. 2008 [978-3-540-79435-6]

Vol. 39: Lucchesi, M., Padovani, C., Pasquinelli, G., Zani, N.
Masonry Constructions: Mechanical
Models and Numerical Applications
176 p. 2008 [978-3-540-79110-2]

Vol. 38: Marynowski, K.
Dynamics of the Axially Moving Orthotropic Web
140 p. 2008 [978-3-540-78988-8]

Vol. 37: Chaudhary, H., Saha, S.K.
Dynamics and Balancing of Multibody Systems
200 p. 2008 [978-3-540-78178-3]

Vol. 36: Leine, R.I.; van de Wouw, N.
Stability and Convergence of Mechanical Systems
with Unilateral Constraints
250 p. 2008 [978-3-540-76974-3]

Vol. 35: Acary, V.; Brogliato, B.
Numerical Methods for Nonsmooth Dynamical Systems:
Applications in Mechanics and Electronics
545 p. 2008 [978-3-540-75391-9]

Vol. 34: Flores, P.; Ambrósio, J.; Pimenta Claro, J.C.;
Lankarani Hamid M.
Kinematics and Dynamics of Multibody Systems
with Imperfect Joints: Models and Case Studies
186 p. 2008 [978-3-540-74359-0

Vol. 33: Nies ony, A.; Macha, E.
Spectral Method in Multiaxial Random Fatigue
146 p. 2007 [978-3-540-73822-0]

Vol. 32: Bardzokas, D.I.; Filshtinsky, M.L.;
Filshtinsky, L.A. (Eds.)
Mathematical Methods in Electro-Magneto-Elasticity
530 p. 2007 [978-3-540-71030-1]

Vol. 31: Lehmann, L. (Ed.)
Wave Propagation in Infinite Domains
186 p. 2007 [978-3-540-71108-7]

Vol. 30: Stupkiewicz, S. (Ed.)
Micromechanics of Contact and Interphase Layers
206 p. 2006 [978-3-540-49716-5]

Vol. 29: Schanz, M.; Steinbach, O. (Eds.)
Boundary Element Analysis
571 p. 2006 [978-3-540-47465-4]

Vol. 28: Helmig, R.; Mielke, A.; Wohlmuth, B.I. (Eds.)
Multifield Problems in Solid and Fluid Mechanics
571 p. 2006 [978-3-540-34959-4

Elastomere Friction

Theory, Experiment and Simulation

Dieter Besdo, Bodo Heimann, Manfred Klüppel,
Matthias Kröger, Peter Wriggers, and
Udo Nackenhorst

Prof. Dieter Besdo
Institut für Kontinuumsmechanik
Gottfried Wilhelm Leibniz Universität
Appelstr. 11
30167 Hannover
Germany

Prof. Bodo Heimann
Gottfried Wilhelm Leibniz Universität
Institut für Mechatronische Systeme
Appelstraße 11a
30167 Hannover
Germany

Priv. Manfred Klüppel
Deutsches Institut für
Kautschuktechnologie e.V.
Eupener Str. 33
30519 Hannover
Germany

Prof. Matthias Kröger
Institut für Maschinenelemente
Konstruktion und Fertigung
Technische Universität Bergakademie
Freiberg, Lampadiusstr. 4
09596 Freiberg/Sachsen
Germany

Prof. Udo Nackenhorst
Institut für Baumechanik und
Numerische Mechanik
Gottfried Wilhelm Leibniz Universität
Appelstr. 9A, 30167 Hannover
Germany

Prof. Dr.-Ing. Peter Wriggers
Institut für Kontinuumsmechanik
Gottfried Wilhelm Leibniz Universität
Appelstr. 11
30167 Hannover
Germany

ISBN: 978-3-642-10656-9   e-ISBN: 978-3-642-10657-6

DOI 10.1007/ 978-3-642-10657-6

Lecture Notes in Applied and Computational Mechanics   ISSN 1613-7736

e-ISSN 1860-0816

Library of Congress Control Number: 2010921208

© Springer-Verlag Berlin Heidelberg 2010

This work is subject to copyright. All rights are reserved, whether the whole or part of the material is concerned, specifically the rights of translation, reprinting, reuse of illustrations, recitation, broadcasting, reproduction on microfilm or in any other ways, and storage in data banks. Duplication of this publication or parts thereof is permitted only under the provisions of the German Copyright Law of September 9, 1965, in its current version, and permission for use must always be obtained from Springer. Violations are liable for prosecution under the German Copyright Law.

The use of general descriptive names, registered names, trademarks, etc. in this publication does not imply, even in the absence of a specific statement, that such names are exempt from the relevant protective laws and regulations and therefore free for general use.

Typeset & Cover Design: Scientific Publishing Services Pvt. Ltd., Chennai, India.

Printed on acid-free paper

9 8 7 6 5 4 3 2 1 0

springer.com

Preface

Understanding elastomer friction is essential for the development of tyres, but also for sealings and other components. Thus it is of great technical importance. There are many aspects to modelling frictional processes in which an elastomer is interacting with a rough surface, ranging from theoretical formulations, leading to reduced and complex models, via numerical simulation techniques to experimental investigations and validations.

The following aspects which all contribute to the modelling of elastomer friction are discussed within the next contributions in more detail: Constitutive modelling of a tyre is of great importance since a great part of the friction stems from hysteretic effects. Thus two contributions are concerned with material modelling of elastomers. The first discusses a specific constitutive model for rubber, MORPH, and its refinement by adding thermal effects like changes in dissipation and elastic behaviour with rising temperatures. The results are based on uniaxial tension tests with different amplitudes and temperatures. While this approach can be viewed in the light of a classical constitutive theory, the second approach uses micro-structure based formulations. Here the material parameters relate to physical quantities within the tube model for rubber elasticity including stiff filler clusters.

These models lead to a good desciption of hysteretic rubber behaviour which then can be included in frictional models. The first one discusses the influence of silica filler content in rubber on the friction behaviour for both wet and dry surfaces at different velocities. The experimental investigations as well as numerical simulation are performed based on a friction model for rough fractal surfaces.

The second approach is based on the fact that the elastomer undergoes large deformations during contact with the road. Hence a model is developed for finite deformations. In this contribution a multi-scale approach is used for a better understanding of the sources of elastomer friction in a more rigorous way at finite strains.

These investigations are then applied to different technical applications related to tyres. The first deals with rolling contact, numerically as well as experimentally. For the latter a moving test rig was developed and assembled. This rig measured the resulting friction characteristics of different elastomers in contact with a glass surface. Based on the dynamic behaviour of the rolling contact observed in the measurements, a mechanical model for simulation is presented.

Another contribution provides a finite element model for rolling contact. Here a mathematically consistent theory for the solution of the advection for inelastic constitutive properties is described which is needed to transport the internal variables corresponding to the material particles motion within the spatially fixed finite element mesh. This approach yields a staggered scheme for the investigation of inelastic properties of rolling wheels at finite deformations.

High-frequency stick-slip vibrations of a tread block occur e.g. on a corundum surface within a certain parameter range which are compared to simulation results. These are investigated using a specific model of the rolling process of a tyre where the tread block follows a trajectory which is obtained from the deformation of the tyre belt. Here the results from simulations with a single tread block provide a deeper insight into highly dynamic processes that occur in the contact patch such as tyre squeal, run-in or snap-out effects.

The last contribution deals with rubber friction under wet conditions at low speeds, which is affected by the micro texture. Experimental investigations are performed using the Grosch wheel and several asphalt surface samples. Parameters like temperature, speed, wheel load, rubber compound and pavement roughness are considered in order to derive possible interactions with respect to the friction coefficient.

All contributions and results are the outcome of the research unit FOR 492 funded by the German Science Council (DFG). This support is gratefully acknowledged.

Our colleague Professor Karl Popp passed away during the funding period of FOR 492 on 24 April 2005, after a serious illness. He was the first spokesman of our group and contributed to a large extent to its success with his deep knowledge in experimental and theoretical mechanics. We owe much to him and will always honour his memory.

Hannover,
29 September 2009

P. Wriggers

Contents

Modelling of Dry and Wet Friction of Silica Filled
Elastomers on Self-Affine Road Surfaces 1
L. Busse, A. Le Gal, M. Klüppel

Micromechanics of Internal Friction of Filler Reinforced
Elastomers 27
H. Lorenz, J. Meier, M. Klüppel

Multi-scale Approach for Frictional Contact of Elastomers
on Rough Rigid Surfaces 53
Jana Reinelt, Peter Wriggers

Thermal Effects and Dissipation in a Model of Rubber
Phenomenology 95
D. Besdo, N. Gvozdovskaya, K.H. Oehmen

Finite Element Techniques for Rolling Rubber Wheels 123
U. Nackenhorst, M. Ziefle, A. Suwannachit

Simulation and Experimental Investigations of the Dynamic
Interaction between Tyre Tread Block and Road 165
Patrick Moldenhauer, Matthias Kröger

Micro Texture Characterization and Prognosis of the
Maximum Traction between Grosch Wheel and Asphalt
Surfaces under Wet Conditions 201
Noamen Bouzid, Bodo Heimann

Experimental and Theoretical Investigations on the
Dynamic Contact Behavior of Rolling Rubber Wheels 221
F. Gutzeit, M. Kröger

Author Index 251

Modelling of Dry and Wet Friction of Silica Filled Elastomers on Self-Affine Road Surfaces

L. Busse, A. Le Gal*, and M. Klüppel

Abstract. We investigate the influence of silica filler content in SBR rubber on the friction behaviour on wet and dry surfaces (rough granite and asphalt) at different velocities experimentally and by simulation, using a recently developed friction model for rough fractal surfaces. The wet friction is shown to be related to pure hysteresis effects, whereas the dry friction also involves adhesion, which is traced back to crack opening mechanisms. It is shown that by calculating relaxation time spectra, the number of free fit parameters can be reduced. These fit parameters are found to vary systematically with filler content for both substrates, and a physical explanation is given. Still, the results of simulations can well be adapted to the measurements. Generally, friction increases with filler concentration on wet substrates. The dry (adhesion) friction turns out to establish a high velocity plateau that becomes lower but more pronounced with increasing filler amount. This is in agreement with experimental master curves for the friction coefficient found in literature, and directly related to other simulation output like the decreasing and flattening of the true contact area with increasing filler contents.

1 Introduction

Friction is a crucial part in every mechanical process. It can be useful or unwanted – but in order to either increase or decrease friction in a system we need to understand the interaction of the parameters that rule this dynamic phenomenon, especially the interaction of material properties, surface properties and lubricant. This can be achieved by referring to the fractal nature

---

L. Busse · A. Le Gal · M. Klüppel
Deutsches Institut für Kautschuktechnologie e.V., Eupener Straße 33,
D-30519 Hannover, Germany
e-mail: manfred.klueppel@dikautschuk.de

* Current address: Ciba Inc., CH-4002 Basel, Switzerland.

of many rough surfaces, allowing for a mathematical description of friction phenomena [1–13].

In this paper, we investigate the role of different amounts of filler in elastomers in respect to their friction behaviour on wet and dry rough granite and asphalt surfaces, respectively. Samples have been filled with silica ranging from no filler to 80 phr silica. Their viscoelastic properties are measured and relaxation time spectra are evaluated. Surfaces profiles are measured and described with few statistical fractal parameters to simulate various functions. Apart from the friction coefficient, other simulation data is achieved as well.

As a refinement of former techniques, we use a bifractal description of both substrate surfaces in order to better approximate the influence of surface structures on the friction process. Additionally, we improve the model of the dry friction by applying experimental results as material fit parameters instead of free fit parameters.

2 Theory

2.1 Analysis of Self-Affine Surfaces

In order to take the properties of the rough substrate surface into account, we need to find a statistical description for its most prominent features. Luckily, most standard surfaces, and especially the granite and asphalt we used, turn out to have a self-affine behaviour [1]. This means that the substrate looks qualitatively the same at different magnifications α in the xy-plane and α^H in z-direction with the Hurst coefficient H. It is connected to the fractal dimension D by

$$D = 3 - H \tag{1}$$

for a 3-dimensional embedding space. The fractal dimension ranges from 2 to 3 for rough surfaces.

The self-affine character of the surface holds up to a cut-off length which differs in a horizontal lateral and a vertical direction. We denote the lateral part as $\xi_\parallel$ and the vertical part as $\xi_\perp$.

In order to find out which cut-off lengths apply to a given surface and to achieve a value for the fractal dimension, we calculate the height-difference correlation (HDC) function $C_z(\lambda)$.

$$C_z(\lambda) = \langle (z(x + \lambda) - z(x))^2 \rangle. \tag{2}$$

The HDC is a measure of how strongly neighbouring points a related to each other. Assuming two points of the surface have the distance λ, their heights are $z(x)$ and $z(x + \lambda)$, respectively. The squares are averaged using the "$\langle \ \rangle$" average over all realizations of the rough surface. The profile does not need to be symmetric [2].

In many cases the HDC graph is split into three regimes of rough surfaces: Above the cut-off length $\xi_\parallel$ the surface is flat, i.e. no correlation between the points can be found except for a similar common height level that determines $\xi_\perp$; this shall be called here macroscopic range. Below the $\xi_\parallel$ down to the lowest measurable length λ_c the HDC can be described well by two linear functions and thus can be approximated well by two scaling ranges. As a result, two different fractal dimensions D_1 and D_2 occur, meeting at the cross over length λ_2, separating the microscopic and mesoscopic range. The relationship between C_z and the surface parameters $\xi_\perp$, $\xi_\parallel$, λ_2, D_1 and D_2, which we can extract from the HDC, is the following for the "mesoscopic" range at $\lambda_2 < \lambda < \xi_\parallel$ (see [3]):

$$C_z(\lambda) = \xi_\perp^2 \cdot \left(\frac{\lambda}{\xi_\parallel}\right)^{2H_1} \tag{3}$$

and for the "microscopic" range at $\lambda < \lambda_2$

$$C_z(\lambda) = \xi_\perp^2 \cdot \left(\frac{\lambda}{\lambda_2}\right)^{2H_1} \cdot \left(\frac{\lambda_2}{\xi_\parallel}\right)^{2H_1}. \tag{4}$$

By Fourier transformation to the frequency space with $\omega_{\min} = 2\pi\nu/\xi_\parallel$ and $\omega_2 = 2\pi\nu/\lambda_2$ where ν is the sample velocity, the corresponding power spectrum density $S(\omega)$ can also be written for the two scaling ranges [4]:

$$S_1(\omega) = \frac{(3-D_1)\cdot \xi_\perp^2}{2\pi\nu\xi_\parallel} \cdot \left(\frac{\omega}{\omega_{\min}}\right)^{-\beta_1} \tag{5}$$

for $\omega_{\min} < \omega < \omega_2$ or

$$S_2(\omega) = \frac{(3-D_1)\cdot \xi_\perp^2}{2\pi\nu\xi_\parallel} \cdot \left(\frac{\omega_{\min}}{\omega_c}\right)^{\beta_1} \cdot \left(\frac{\omega}{\omega_c}\right)^{-\beta_2} \tag{6}$$

for $\omega_2 < \omega$ respectively, where we abbreviate the fractal dimension for $i = 1, 2$ with $\beta_i = 7 - 2D_i$. If more than two scaling ranges should be necessary, the formulas can be expanded to any wanted number of multifractality [4–6].

Not all surface points get in contact with the sliding elastomer. The relevant summit height distribution $\Phi(z_s)$ can be calculated from the height distribution $\Phi(z)$ with its maximum at $z_{\max}$ by an affine transformation with a parameter s as a surface constant:

$$z_s = \frac{(z - z_{\max})}{s} + z_{\max}. \tag{7}$$

This s-parameter is gained by fitting the height distribution of the substrate profile to the summit height distribution which takes into account only the local maxima that have direct contact with the sample [6].

2.2 Hysteresis Friction Simulation

Friction μ_{ges} can be divided into two parts: adhesion friction μ_{Adh} and hysteresis friction μ_{Hys}.

$$\mu_{\text{ges}} = \mu_{\text{Adh}} + \mu_{\text{Hys}}. \tag{8}$$

The latter is caused by the energy dissipations where the rubber sample is deformed at local asperities. The excited frequencies follow a spectrum as given in Equations (5) and (6) and thus can be used for the hysteresis integral [6–9] for the friction force F_{Hys} under normal force F_N and thus friction coefficient μ_{Hys} is according to our model

$$\mu_{\text{Hys}}(\nu) \equiv \frac{F_{\text{Hys}}}{F_N} \tag{9}$$

$$= \frac{\langle \delta \rangle}{2\sigma_0 \nu} \left(\int_{\omega_{\min}}^{\omega_2} \omega \cdot E''(\omega) \cdot S_1(\omega) d\omega + \int_{\omega_2}^{\omega_{\max}} \omega \cdot E''(\omega) \cdot S_2(\omega) d\omega \right).$$

Again we assume a bifractal approach to suit best the nature of the HDC. Microstructures influence mainly to the lower velocities [6]. E'' is the loss modulus of the elastomer, σ_0 is the applied pressure and $\langle \delta \rangle$ is the mean excitation depth inside the rubber

$$\langle \delta \rangle = b \cdot \langle z_p \rangle \tag{10}$$

with the mean penetration depth z_p of the asperities into the rubber, scaled by the factor b. Further, only wave lengths above $\lambda_{\min} = 2\pi\nu/\omega_{\max}$ contribute to Equation (9). It holds that [8]

$$\frac{\lambda_{\min}}{\xi_\parallel} \cong \left(\left(\frac{\lambda_2}{\xi_\parallel} \right)^{3(D-2-D_1)} \frac{0.09\pi \xi_\perp \cdot |E(2\pi\nu/\lambda_{\min})| \cdot F_0(t) \cdot 6\pi \cdot \sqrt{3}\lambda_c^2 \cdot n_s}{s^{2/3} \cdot \xi_\parallel \cdot |E(2\pi\nu/\xi_\parallel)| \cdot F_{3/2}(t_s)} \right)^{\frac{1}{3D_2-6}}, \tag{11}$$

where $n_s \sim (3-\beta_2)/(5-\beta_2)$ is the summit density and

$$F_n = \int_t^\infty (x-t)^n \cdot \phi(x) dx \tag{12}$$

with $n = 0, 1, 3/2$ are the Greenwood–Williams functions [10] with the normalized distances $t = d/\sigma_{\text{HD}}$ and $t_s = d/\sigma_{\text{SHD}}$, where the gap distance d indicates the rubber distance from the mean substrate level; σ_{HD} and σ_{SHD} are the standard deviations of the height distribution or summit height distribution, respectively.

For pure hysteresis friction, it becomes [7, 9]

$$\lambda_{\min} \cong \sqrt{\frac{|E(\lambda_{\min}) \cdot C_z|(\lambda_{\min})}{\sigma(\lambda_{\min})}}. \tag{13}$$

Of great importance is the true contact area A_c, which is calculated as a ratio to the nominal contact area A_0 as [5]

$$\frac{A_c}{A_0} \approx \left(\frac{\xi_\perp \cdot F_0^2(t) \cdot F_{3/2}(t_s) \cdot |E(2\pi\nu/\xi_\parallel)|\tilde{n}_s^2}{808\pi \cdot s^{3/2} \cdot \xi_\perp \cdot |E(2\pi\nu/\lambda_{\min})|} \right)^{1/3}. \quad (14)$$

2.3 Adhesion Friction Fitting

While hysteresis friction is sufficient to describe tribologic effects on typical wet contacts, adhesion has to be considered additionally in only partly lubricated [11], and especially on dry systems, because direct contact between rubber and substrate allows molecular interactions with the force F_{Adh}. This leads to the adhesion friction coefficient [9, 12, 13]

$$\mu_{\text{Adh}} = \frac{F_{\text{Adh}}}{F_N} = \frac{\tau_s \cdot A_c}{\sigma_0 \cdot A_0}, \quad (15)$$

where the load σ_0 is the applied pressure, whereas the velocity dependent interfacial shear stress τ_s produces peeling effects of the rubber front side at local asperities causing crack propagation, and can be written as [14]

$$\tau_s = \tau_0 \cdot \left(1 + \frac{E_\infty/E_0}{(1 + (v_c/v))^n} \right). \quad (16)$$

Obviously, we find a dependency on velocity. The ratio of mechanical moduli as step height for the glass transition is found experimentally (see Section 4.1). The parameters $\tau_0 = \tau_s$ ($v = 0$) and the critical velocity v_c are subject of free fitting, for practical reasons. The latter can be interpreted as the point where the shear stress converges to a maximum [15]: $\tau_s(v_c) \approx \tau_s(\infty)$. The exponent n [16]

$$n = \frac{1 - m}{2 - m} \quad (17)$$

is gained from the power law exponent $m(\tau) < 1$ of the relaxation time spectra $H(\tau)$ [17] in the glass transition range as iterative approximation

$$H(\tau) = A \cdot G' \cdot \frac{d \log(G')}{d \log(\omega)} \cong \tau^{-m} \quad (18)$$

where the relaxation time spectra can be evaluated from frequency dependent master curves of the storage modulus G' with the relaxation time $\tau = 1/\omega$, not to be confused with the shear strength τ_s, by using the correctional term

$$A = (2 - \alpha)/2\Gamma \left(2 - \frac{\alpha}{2} \right) \cdot \Gamma \left(1 + \frac{\alpha}{2} \right) \quad (19)$$

applying the local slope α to the gamma function Γ.

3 Experimental Methods and Proceedings

3.1 Surface Properties

As a first step, substrates were analyzed due to their surface properties. Profiles were taken with stylus measurements for granite, and white light interferometry for asphalt, respectively.

The single lines $z(x)$ were transformed to zero level and zero slope, then averaged and summarized to height distributions $\Phi(z)$ (Figure 1). Although both granite and asphalt have almost Gaussian distributions, asphalt is a bit asymmetric and broader, which means larger height differences and thus more vertical roughness.

According to Equation (2) we calculated the HDC functions, which is shown in Figure 2 based on profile scans of the granite and asphalt surfaces we used for our experiments and simulations. Both curves can be described by two linear interpolations below their cut-offs, giving the fractal dimensions for their scaling ranges. The cut-off lengths are also visible. All surface parameters after bifractal analysis are summarized in Table 1.

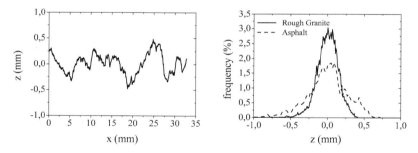

Fig. 1 Height profile of granite (left) and the resulting height distribution, compared to asphalt (right).

Table 1 Bifractal surface descriptors and affine parameter s for granite and asphalt.

Surface Descriptors	Granite	Asphalt
D_1	2.37	2.39
D_2	2.14	2.09
$\xi_\perp$ [μm]	310	430
$\xi_\parallel$ [μm]	2490	1440
$\xi_\parallel/\xi_\perp$	7.96	3.35
λ_2 [μm]	93,0	332
s-Parameter	1.251	1.250

Modelling of Dry and Wet Friction of Silica Filled Elastomers

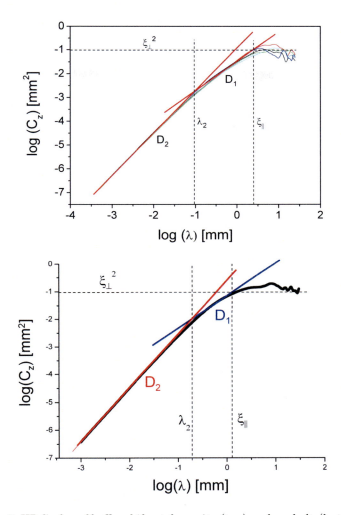

Fig. 2 HDC of a self-affine bifractal granite (top) and asphalt (bottom).

Both surfaces have similar fractal dimensions, dropping to almost flat in the mesoscopic range. Their vertical roughnesses are only slightly different, too, asphalt being a bit rougher. The horizontal cut-off length, on the other hand, is quite different, leading to a much higher parallel-orthogonal ratio on granite: the surface is less sharp. Finally, the microscopic range lasts on asphalt to larger lengths. This is due to the grain size that is distributed in an especially small interval on asphalt.

Table 2 Recipes of the sample pool.

Sample	SBR 2525	Silica Φ	Silan [Si_{69}]	ZnO	Ste. Acid	IPPD	CBS	S
S0K	100	0	–	3	1	1.5	2.5	1.7
S2K	100	20	1.7	3	1	1.5	2.5	1.7
S4K	100	40	3.3	3	1	1.5	2.5	1.7
S6K	100	60	5	3	1	1.5	2.5	1.7
S8K	100	80	6.7	3	1	1.5	2.5	1.7

3.2 Material Preparation and Properties

The elastomer samples S-SBR 2525 (25% styrol, 25% vinyl) as the other friction partner were prepared as given in Table 2 with a thickness of 2 mm. The number in the sample name denotes the amount Φ of silica filler (Ultrasil GR7000, Evonik) in 10 phr units.

All samples have been investigated in their viscoelastic properties by dynamic mechanical analysis (DMA) on samples with a $10 * 30$ mm rectangular geometry. Temperature sweeps and frequency sweeps between -50 and $+60°C$ at 5 to $10°C$ steps with an amplitude of 0.5% have been done and extracted into master curves for the shear moduli G' and G''. A polynomial fit of these results was inserted into the Maple script for the simulation.

To confirm the shift factors for the master curves, dielectrical measurements have been done [18] over a range from -100 to $+100°C$ in discrete steps of $5°C$, while varying the frequency. The sample geometry is a gold sputtered circle of 40 mm at a fixed force.

3.3 Friction Experiments and Simulations

Friction experiments on dry and wet surfaces with a pressure of 12.3 kPa on the $50*50$ mm samples have been performed at room temperature in order to verify the simulations. The velocity ranges from 0.01 up to 30 mm/s and has been held stationary, sliding the elastomer fixed with a wire over the surface (Figure 3). All velocities have been examined several times, from fast to slow and again the other way. The wet friction, done with a 5%vol tenside in water solution covering the substrate, gives pure hysteresis effects, whereas the dry friction also contains the adhesion part.

Containing all necessary data, simulations have been conducted for all relevant combinations of filler, elastomer and substrate. This results in the velocity dependent hysteresis friction coefficient and other friction data, like the real area of contact, which shall be discussed in Section 4.4. Simulations are gained from a Maple script embedding the sample/surface data into the DIK model, and then have to be fit to the measurements.

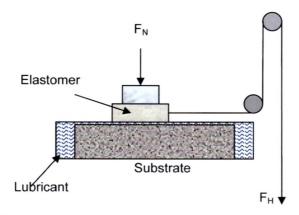

Fig. 3 Principle of the sliding friction experiment.

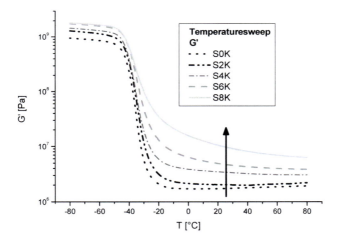

Fig. 4 T-sweeps offer the temperature dependent modulus G'.

4 Results and Discussion

4.1 Viscoelastic Properties

Dynamic mechanical analysis reveals how the viscoelastic properties depend on the filler content. In the T-sweeps (Figure 4), the filler increases the moduli monotonously, especially for high temperatures, so the ratio $E'_\infty/E'_0 = G'_\infty/G'_0$ decreases (see Table 1 for values). This behaviour plays an important role in the adhesion friction. Further, the glass transition is broadened considerably compared to that of unfilled S0K, because chain mobility is reduced in the presence of filler, as can be seen in the curves of G'. The

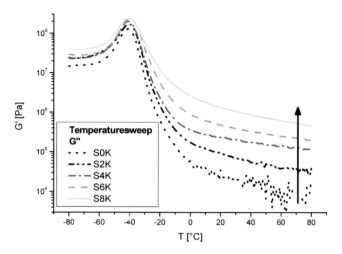

Fig. 5 T-sweeps for the loss modulus G''.

effect is minimal for weakly filled samples (S2K), moderate for medium filled ones (S4K, S6K) and dominant for highly filled sample S8K.

Looking at the loss modulus G'' in Figure 5, the increase with filler content is obvious, too, but more proportional to the filler amount than in the T-sweep of G'. In both cases, the increase is smaller for low temperatures, as chain mobility is limited anyway under these circumstances.

According to the time-temperature-superposition principle (TTS), temperature influence can be replaced by frequency influence, where cold rubber equals high frequency treatment and causes a glassy state with high stiffness, whereas a warm elastomer dominated by rubber matrix/ filler is connected to low frequencies. Indeed, by applying various frequencies at discrete temperatures on the sample, we get a bunch of branches for the moduli that can be constructed to a master curve for the whole frequency range over several decades by shifting the branches horizontally in order to form a continuous curve. Above the glass transition temperature, the shift factors obey the WLF relationship as shown in Figure 6, with WLF constants $C_1 = -3.85$ and $C_2 = 91.2°C$ at $T_{ref} = 20°C$ to match the room temperature of friction measurements. Below about $-45°C$, the relative shift between the branches decreases.

These horizontal factors were extracted from the unfilled sample S0K (Figure 7), but are valid also for the filled samples. For high temperatures, the branches of filled samples are also vertically dispersed, so vertical shift factors need to be applied as well, independently for G' and G''. The TTS is not valid anymore for filled elastomers in this temperature range, because the filler-filler bond instead of the polymer matrix then dominates the viscoelastic behaviour [19]. Shift factors for $\log G'$ and $\log G''$ with linear slope in the

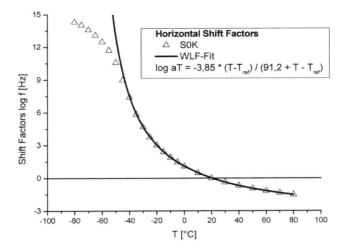

Fig. 6 Horizontal shift factors for the master curves.

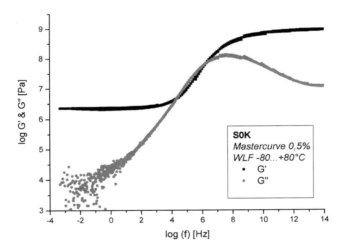

Fig. 7 Master curve for unfilled S0K.

corresponding inverse temperature range (an Arrhenius plot) are granted in Figure 8. The shift of the loss modulus depends more on the temperature than the one for the storage modulus. This is easy to see in Figure 9, which shows an example of a master curve that is shifted horizontally but not vertically.

An overview of master curves in Figures 10 and 11 gives similar results as for the T-sweeps. Again both moduli G' and G'' are increased at high temperature (corresponding to low frequency) by adding more filler, whereas the high frequency range reflects only minor change with filler amount. One

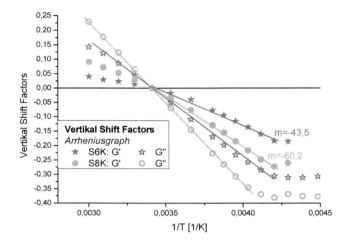

Fig. 8 Vertical shift factors Arrhenius plot.

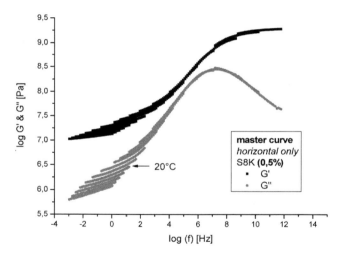

Fig. 9 Horizontal master curve for S8K before vertical shifting, 20°C branch indicated.

remarkable feature is the maximum of the loss modulus G'', which does not change its frequency with the filler amount and displays nearly the same value at this point for every sample.

As a result, the frequency of the maximum for $\tan \delta = G''/G'$ does not change either with filler content, but as G'' stays the same while G' increases, the value of the maximum decreases strongly when filler is added (Figure 12). The frequency dependency of G' and G'' becomes not only smaller but also more similar with rising filler amount.

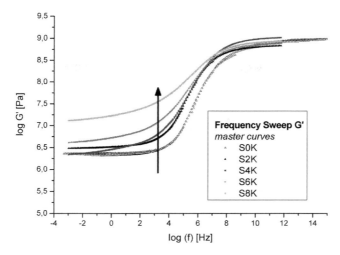

Fig. 10 Frequency dependent G' constructed from f-sweeps.

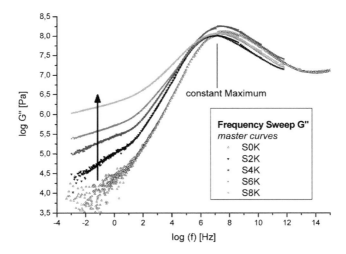

Fig. 11 G'' master curves from f-sweeps.

Another relationship that can be gained from the DMA data are relaxation time spectra. By calculating the relaxation time spectra with (18) we find values for the fitting exponent n from taking the steepest negative slope of the Ferry method with two iteration steps. The condition $m < 1$ turns out to be fulfilled in our case. In Figure 13 we find that the most significant part of the relaxation spectrum, the slope m between absolute maximum at $\tau < 10^{-7}$ s

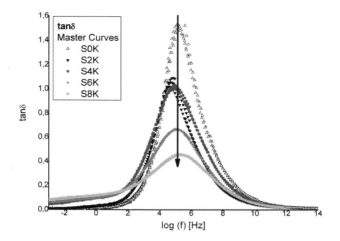

Fig. 12 $\tan\delta$ taken from f-sweeps.

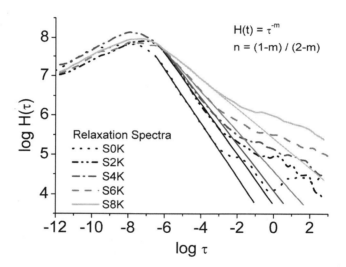

Fig. 13 Relaxation time spectra with slope lines for exponents.

and beginning minimum at $\tau > 10^{-3}$ s, is flattening more and more by increasing the filler, resulting in increasing n as given in Equation (17). Values for this are noted in Table 3. Notice that m is defined as negative exponent, so it takes positive values on the descending flank of the spectra, leading to positive values for n as well. We will use these exponents to fit simulation to adhesion friction measurements in Section 4.3.

Table 3 Fixed, surface independent fit parameters.

Sample	E_∞/E_0 (T-sweep)	E_∞/E_0 (f-sweep)	Slope m	Exponent n
S0K	491.2	473.2	0.714	0.209
S2K	584.3	298.2	0.711	0.211
S4K	475.2	541.1	0.594	0.283
S6K	457.7	389.2	0.441	0.357
S8K	291.7	177.8	0.358	0.390

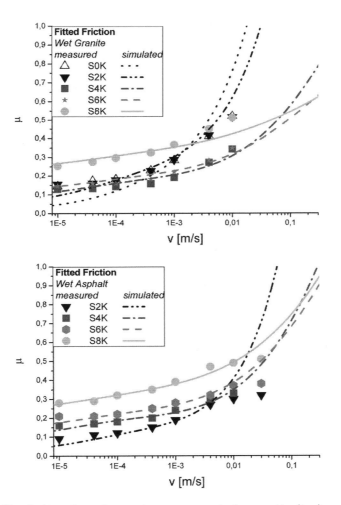

Fig. 14 Simulation adapted to wet measurements for granite (top) and asphalt (bottom).

4.2 Friction Measurements

In order to investigate the pure hysteresis friction, it is necessary to extinguish adhesion friction by separating rubber from substrate so no direct molecular interaction is possible. This can be realized by applying a low viscosity lubricant film. The results of these wet measurements are shown in Figure 14. On both granite and asphalt surfaces, friction clearly increases with velocity, which means lubrication is effective especially at lower speeds. In case of granite the steep slope at 10 mm/s indicates a further strong increase of friction, whereas for asphalt higher velocities mean only a slight increase of friction. This is due to the lateral scaling parameter $\xi_\parallel$ (see Table 3): asphalt, which is laterally smoother than granite, exhibits the beginning of the plateau area at lower velocities. The effect of temperature can be neglected especially on wet surfaces for most velocities in the range examined here, but might play a role for high velocities, especially without lubricant [20].

The amount of friction is similar for both substrates. In general, friction on granite tends to be slightly higher than on asphalt. The exact behaviour strongly depends on the filler amount: For asphalt, the friction increases monotonously with the degree of filler in the complete velocity range. On granite, this is true only when disregarding weakly filled samples (S0K, S2K). This increase of friction can be explained by the increased hysteresis of filled elastomers (see Figure 12).

The total friction μ_{ges} (Figure 15) in the dry systems is drastically increased by the presence of adhesion, which is $\Delta\mu = \mu_{\text{Adh}} = \mu_{\text{ges}} - \mu_{\text{Hys}}$. Filler does not influence the amount of friction as much as in the hysteresis case, but effects strongly the shape of the friction curves: All adhesion curves display a maximum because the true area of contact, which plays a vital role for adhesion (see Section 4.4), decreases with velocity. This behaviour will be explained in Section 4.3 when comparing it to simulations. Apart from high velocities, where weakly filled samples display an almost constant plateau that is only slightly diminished by more filler, adhesion friction decreases generally with filler amount in all systems, also for low velocities. This can be explained by the higher hardness of filled elastomers, which prevents the rubber from intense contact and thus from reaching large true contact areas. Summing up, the dry friction is as well decreased with filler, though not as strictly as for pure adhesion. Again, both substrates own similar friction curves. Friction coefficients are almost identical for low filler samples and rather similar for high filler samples at low velocities, which are increasing less with velocity for measurements on asphalt, making the curves more strictly related directly to filler amount. With a larger $\xi_\parallel/\xi_\perp$ ratio, granite offers larger gripping even for filled samples at high velocities, which will be confirmed in Section 4.4 when regarding the true contact areas. This explains why highly filled samples have a bit higher friction at fast sliding.

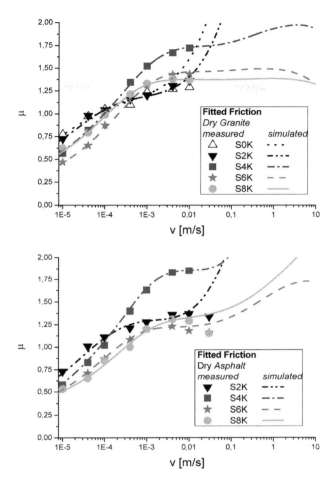

Fig. 15 Fitted adhesion simulation matches dry friction for granite (top) and asphalt (bottom).

4.3 Adapting Friction Simulation to Wet and Dry Measurements

Next we will see how well our experiments can be described or even be predicted by simulations. As a first step, we compare the simulated hysteresis friction scaled with factor b (Table 4) to the data of wet friction measurements. The values for b start at high numbers for unfilled samples, then quickly decrease with filler amount and stay almost constant above the percolation threshold at 40 phr silica. On granite, b is higher than on asphalt. The relation of this fit factor is given in Equation (10). This means that the effect penetrating asperities propagates more deeply in the unfilled samples

Table 4 Free fit parameters for hysteresis and adhesion simulation.

Sample	Granite			Asphalt		
	b	τ_0	v_c	b	τ_0	v_c
	–	kPa	mm/s	–	kPa	mm/s
S0K	70	5.91	0.020	–	–	–
S2K	27	6.56	0.045	10	64	0.62
S4K	7	15.4	1.90	4.5	150	9.00
S6K	6.8	18.8	0.90	5.5	110	0.60
S8K	7.5	48.8	0.33	7	420	1.00

up to 40 phr. The simulation is shown in Figure 14 as drawn lines for all samples, both on granite (top) and asphalt (bottom). Obviously, the measured curves can be reproduced fairly well with simulation for low and moderate velocities, but is less accurate when sliding faster than a few mm/s: Then the simulations are much steeper than measurements for low fillings (S0K, S2K), more or less correct for medium fillings (S4K, S6K) and not steep enough for high fillings (S8K), on a granite surface. For asphalt, simulations behave the same way but are slightly steeper than on granite in comparison to the measurements, and fit the measurements still at higher velocity. The answer can be found in the surface descriptors (Table 1): as $\xi_\parallel$ indicates, granite is rougher than asphalt on the lateral scale, so the simulation will be valid to higher velocities. The highly filled simulated curves are less steep and thus less dependent on velocity because their elasticity depends less on temperature and thus less on frequency than the weakly filled samples (see Section 4.1).

The second step is to add an adhesion term as explained in Section 2.3 to the hysteresis fit. For the best combinations of the free and given fit parameters the simulation are compared to the dry measurements in Figure 15. Apart from the b parameters for hysteresis friction, more fit parameters were necessary for dry friction. Some of them could directly be taken from the results of rubber material experiments, and do thus not depend on the contacting surface: the ratio E_∞/E_0 was measured as part of the DMA and the exponent n was extracted from relaxation time spectra as described before. Two free fit parameters, v_c and τ_0, were added to gain the total friction completely, individually for each surface. All surface dependent fit parameters including b, which stays unchanged, are displayed in Table 4. The shear stress rises with filler amount for both granite and asphalt, especially for highest filler content. Simulation range and accuracy of the simulation are only limited by knowledge of the regarded viscoelastic and surface properties. The critical velocity displays a maximum at 40 phr for both substrates.

Again, the fit is excellent for most points, even at higher velocities on granite, but fails to mirror the fast velocities on asphalt. While temperature effects by friction heating could be assumed to be neglectable on wet surfaces, they may indeed play a role on dry substrates when the sliding speed is high enough. The heating results in an increasing elasticity, which means the hysteresis part for dry friction is lower than for wet friction. Hysteresis friction is still increasing with velocity, but its steepness is reduced by filled amount already on wet surfaces. Combined with the decreasing real area of contact at high velocities, dry friction may be reduced in effect, as seen for asphalt when sliding fast. As a result of these two contrary effects, dry friction enters a plateau for velocities above some mm/s, as the wet part is still increasing. This behaviour has already been found by Grosch [21] as master curves of samples filled with 50 phr highly active carbon blacks (N220, N330) sliding on dry clean silicone and in [6] for carbon black filled S-SBR5025 on rough granite. In our case, the same behaviour is found for the simulations of highly silica filled systems: On granite, dry friction follows clearly this plateau shape for the filled samples. On asphalt, the effect is present but less accented. The effect is confirmed by simulation and well visible on the extended velocity scale.

4.4 Contact Simulations

Apart from verifying the DIK model for friction, simulation also gives other interesting results, which shall be presented in this section.

The gap distance d (Figure 16) is a distance between the lowest rubber level and the mean profile height $\langle z \rangle$ of the substrate. It increases with velocity (Figure 17), as the time for the elastomer to enter the cavities of the substrate is reduced. Further, it is clearly visible how filler increases the gap distance. This can be explained by the lower elasticity of filled samples, preventing the rubber to fill the cavities deeply. On asphalt the distance is a bit higher than on granite due to the slightly higher $\xi_\perp$. It is easy to notice the analogy of the filler influence here and in the storage modulus as measured in frequency

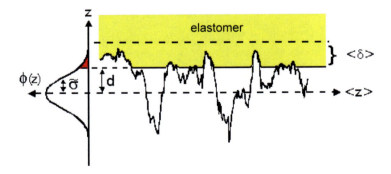

Fig. 16 Definition of the parameters to describe rubber indention.

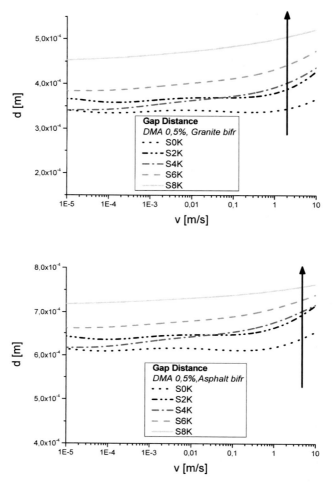

Fig. 17 Simulated gap distance for granite (top) and asphalt (bottom).

sweeps (Figure 10): S2K and S4K are almost indiscriminable, highly filled S8K displays a clearly separated value for all velocities and unfilled S0K drifts apart only for high velocity, which can be interpreted as being caused by the considerably separated curve of unfilled G' at its rising slope up to glass transition. Again, high frequencies are connected to large sliding velocities. We will find the same filler behaviour in all contact simulations, as described next.

Looking at the true contact area in Figure 18, we find a decrease of contact with an increase of filler content, and with velocity. As before, this is caused by the vanishing ability of rubber to fill the gaps and thus get in contact with its interface when either the elasticity or the contact time, at high

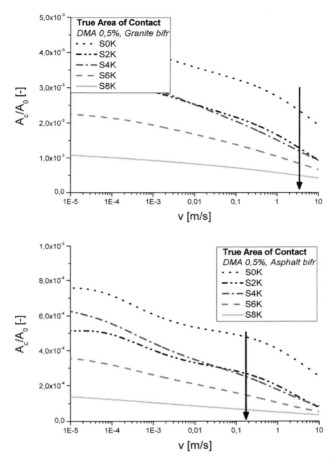

Fig. 18 Simulated true contact area for granite (top) and asphalt (bottom).

velocities, is diminished. Though still in the per mille range, granite offers a larger contact than asphalt. To understand this, we regard not only the vertical, but also lateral surface descriptors: the larger the ratio $\xi_\parallel/\xi_\perp$, the flatter and easier to access are the surface structures, and the more contact is possible. According to Table 1 granite has the flatter surface, explaining the phenomenon. Additionally, this ratio makes contact on granite easier than on asphalt even for high velocities, so the curves for asphalt decrease faster than those for granite.

Part of the adhesion fit as given in Equation (16) is the shear stress $\tau_s(v)$. The curves are displayed in Figure 19 and show how the stress rises with velocity and converges to a plateau of the value $\tau_s(\infty) = \tau_0 * (1 + E_\infty/E_0)$. For very small velocities (not visible in this graph), τ_s becomes the fit parameter τ_0, which is connected to the static surface tension γ_0 by a characteristic

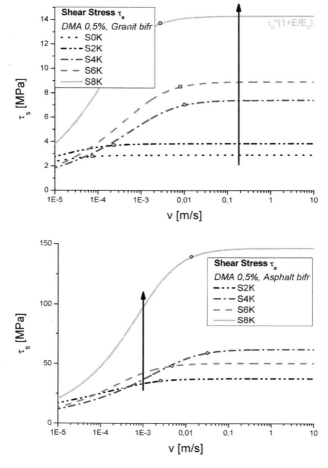

Fig. 19 Best fitted shear stress for granite (top) and asphalt (bottom).

length scale l_s with $\tau_0 = \gamma_0/l_s$. As τ_0 increases at about one decade (Table 4), γ_0 increases slightly only with filler amount, so the l_s, which can be interpreted as a process zone of contact, decreases largely for highly filled samples.

The increase of τ_0 is even large enough to counter the decreasing ratio E_∞/E_0, so the values for τ_s are also rising with filler content – which is understandable because filler decreases the elasticity of the samples – and is significantly higher on asphalt, assumingly due to differences in surface polarity. Again, the filler amounts put the curves into groups of low, moderate and high filler contents, like the T-sweep for G' is presented in Figure 5.

Higher filler amounts also shift the beginning of the plateau to different velocities that reflect the critical velocities $v_c(\Phi)$, reaching a maximum

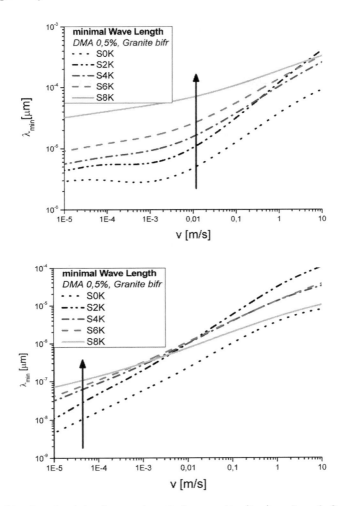

Fig. 20 Simulated minimal wave length for granite (top) and asphalt (bottom) reveal similarity.

already at moderate velocities. The maximum of v_c at 40 phr in Table 4 can be estimated looking at the markers in Figure 19 indicating 95% of the maximum shear stress. The rise of tension starts about one decade later in velocity on asphalt than on granite. The increasing relaxation exponent n makes the curves more distorted with filler amount. This increasing distortion, together with the decreasing E_∞/E_0 and increasing τ_0 are responsible for the v_c reaching its maximum.

The minimal contact length $\lambda_{\min}$ on the lateral direction as defined in (11) is displayed in Figure 20. Higher velocities result in a strongly increasing minimal contact length, which means that at high velocities only large structures contribute to the friction process as fewer small asperities become accessible for the fast sliding rubber.

Higher filler contents also result in higher minimal length, too, because the poorly elastic filled sample cannot enter the cavities deeply to establish a laterally broad contact. For asphalt, this is true only at a second glance: As the highly filled samples have a weaker slope in minimal wave length, curves reverse for high velocities. Looking at smaller velocities reveals the same behaviour as on granite, but on smaller values, as the ratio $\xi_\parallel/\xi_\perp$ is also smaller on asphalt. The slope is explainable by the fact that at high velocities high frequencies are excitated, and the master curves (Section 4.1) show that elasticity depends less on frequency for filled samples.

5 Conclusions

SBR rubber has been filled with silica in amounts of 0, 20, 40, 60, 80 phr. The samples have been tested in their viscoelastic behaviour as well as in their wet and dry friction on granite and asphalt at stationary velocities. The viscoelastic results have been used to simulate the systems using a recently developed model on bifractal substrate surfaces, which is an significant improvement to monofractal description. These simulations have been compared to the measurements as hysteresis friction and fit to adhesion friction.

The shear moduli G' and G'' increase both with filler amount, especially in the high temperature (= low frequency) range. The loss angle $\tan\delta$ decreases, with unchanged frequency of its maximum. Weakly filled samples have a steeper slope in the corresponding relaxation time spectra.

Wet hysteresis friction increases with velocity in all cases, especially on granite. Filled samples possess in general a higher friction, which is more accented on asphalt, though both substrates have similar friction coefficients. Simulation can well describe the wet measurements correctly for low and moderate velocities.

The same is true for the dry friction (hysteresis plus adhesion parts), which starts on a higher level and runs more and more into a plateau for high velocities, most obvious for highly filled samples, although the influence of filler is less pronounced for the general friction coefficient. This plateau is best recognized at very high velocities in the simulations. Replacing free fit parameters by calculated material parameters gained from relaxation time spectra allows to simulate the friction accurately.

The shear stress is maximal for high velocities and filled samples, and higher for asphalt. The gap distance rises with filler content, being larger on asphalt. Complementary, the real area of contact decreases with velocity and filler amount. Asphalt has higher values for the first, granite for the latter function. The minimal contact length increases clearly with velocity and in general with filler amount.

Acknowledgements

The authors thank N. Zhang (Leibniz-University Hannover), who prepared his master thesis at the DIK. This work has been supported by the Deutsche Forschungsgemeinschaft (DFG) as a part of the collaborate research unit DFG-Forschergruppe "Dynamische Konktaktprobleme mit Reibung bei Elastomeren" (FOR 492). This is gratefully acknowledged.

References

1. Mandelbrot, B.: The Fractal Geometry of Nature. W.H. Freeman, New York (1982)
2. Heinrich, G., Klüppel, M.: Rubber friction, tread deformation and tire traction. Wear 265, 1052–1060 (2008)
3. Le Gal, A., Yang, X., Klüppel, M.: Sliding friction and contact mechanics of elastomers on rough surfaces. In: Wriggers, P., Nackenhorst, U. (eds.) Analysis and Simulation of Contact Problems. LNACM, vol. 27, p. 253. Springer, Heidelberg (2006)
4. Meyer, T., Busse, L., Le Gal, A., Klüppel, M.: Simulations of hysteresis friction and temperature effects between elastomers and rough or microscopically smooth interfaces. In: Boukamel, A., et al. (eds.) Constitutive Models for Rubber V, pp. 351–355. Tabor & Francis Group, London (2008)
5. Le Gal, A., Guy, L., Orange, G., Bomal, Y., Klüppel, M.: Modelling of sliding friction for carbon black and silica filled elastomers on road tracks. Wear 264, 606 (2008)
6. Le Gal, A., Klüppel, M.: Investigation and modelling of rubber stationary friction on rough surfaces. J. Phys. Condens. Matter 20, 015007 (2008)
7. Klüppel, M., Heinrich, G.: Rubber friction on self-affine road tracks. Rubber Chem. Technol. 73, 578 (2000)
8. Müller, A., Schramm, J., Klüppel, M.: Ein neues Modell der Hysteresereibung von Elastomeren auf fraktalen Oberflächen. Kautschuk Gummi Kunstst. 55, 432 (2002)
9. Le Gal, A.: Investigation and modelling of rubber stationary friction on rough surfaces, PhD Thesis, University of Hannover (2007)
10. Greenwood, J., Williamson, J.: Contact of nominally flat surfaces. Proc. R. Soc. London A 295, 300 (1966)
11. Kummer, H.: Lubricated friction of rubber discussion. Rubber Chem. Technol. 41, 895 (1968)
12. Le Gal, A., Klüppel, M.: Modelling of rubber friction: A quantitative description of the hysteresis and adhesion contribution. In: Austrell, P.E., Kari, L. (eds.) Constitutive Models for Rubber IV, p. 509. Tabor & Francis Group, London (2005)
13. Le Gal, A., Yang, X., Klüppel, M.: Evaluation of sliding friction and contact mechanics of elastomers based on dynamic mechanical analysis. J. Chem. Phys. 123, 014704 (2005)
14. Le Gal, A., Klüppel, M.: Investigation and modelling of adhesion friction on rough surfaces. Kautschuk Gummi Kunstst. 59, 308 (2006)
15. De Gennes, P.: Soft adhesives. Langmuir 12, 4497 (1996)

16. Persson, B., Brener, E.: Crack propagation in viscoelastic solids. Phys. Rev. E 71, 036123 (2005)
17. Williams, M., Ferry, J.: Second approximation calculations of mechanical and electrical relaxation and retardation distributions. J. Polym. Sci. 11, 169 (1953)
18. Fritzsche, J., Klüppel, M.: Filler networking and reinforcement of carbon black filled styrene butadiene rubber. In: Kautschuk Herbst Kolloquium, Hannover (2008)
19. Klüppel, M.: Evaluation of viscoelastic master curves of filled elastomers and applications to fracture mechanics. J. Phys. Condens. Matter 21, 035104 (2009)
20. Persson, B.: Rubber friction: Role of the flash temperature. J. Phys. Condens. Matter 18, 7789 (2006)
21. Grosch, K.: The relation between the friction and visco-elastic properties of rubber. Proc. R. Soc. London A 274, 21 (1963)

Micromechanics of Internal Friction of Filler Reinforced Elastomers

H. Lorenz, J. Meier*, and M. Klüppel

Abstract. To calculate the mechanical behavior of a loaded component the engineer needs a model which describes the stress-strain-behaviour. In order to relate the properties of rubber and filler to those of a filled elastomer we use a microstructure-based approach where material parameters are physical quantities, instead of mere fit parameters. Core of the model is the hydrodynamic reinforcement of rubber elasticity (tube model) by stiff filler clusters. The deformation is concentrated at a smaller part of the total volume, resulting in an amplification of stress. Under stress, clusters can break and become soft, leading to deformation of larger parts of the volume and related stress softening. The effect is expressed as an integral over the "surviving" section of the cluster size distribution. On the other hand, cyclic breakdown an re-agglomeration of soft clusters causes hysteresis. Filled elastomers also show a certain inelastic behaviour called setting. The corresponding stress contribution is modeled by a semi-empirical dependency with respect to maximum deformation. Using dumbbell specimens, we have done uniaxial stress-strain measurements in tension an compression. Parameterfits show that the model satisfactorily describes compression and tension-tests. Generally, the parameters lie in a physically reasonable range. For the first time hydrodynamic reinforcement (formulated by a reinforcement exponent) and stress softening have been implemented into FE code. A rolling rubber wheel under load is simulated.

1 Introduction

Nanoscopic filler materials like carbon black or silica play an important role in the reinforcement of elastomers. Besides making the elastomer stiffer and

---

H. Lorenz · J. Meier · M. Klüppel
Deutsches Institut für Kautschuktechnologie e.V., Eupener Straße 33,
D-30519 Hannover, Germany
e-mail: `manfred.klueppel@dikautschuk.de`

* Henniges Automotive, Am Buchholz 4, D-31547 Rehburg-Loccum, Germany.

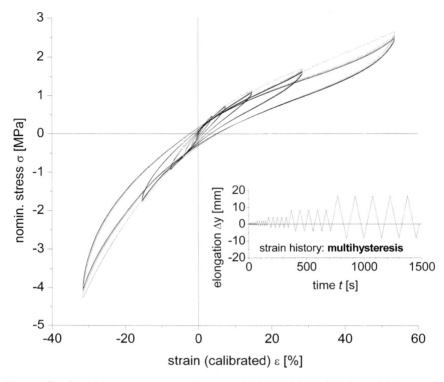

Fig. 1 Combined compression-tension test of CB-filled SBR/BR in multi-hysteresis test; the inset shows the strain history.

tougher, the incorporation of a filler brings about a non-linear dynamic-mechanical response reflected by the amplitude dependence of the dynamic moduli. This effect was investigated by several authors such as Payne [1] and Medalia [2]. A related effect is the stress softening under quasi-static cyclic deformation, which was studied by Mullins intensively [1]. An example of a uniaxially loaded specimen is shown in Figure 1. A drop in stress usually occurs after the loading history has gone beyond the previous maximum. Most of the stress drop at a certain strain occurs in the first cycle, and in the following cycles the specimen approaches a steady state stress-strain curve. A second characteristic effect caused by fillers is the pronounced hysteresis which is related to the dissipation of mechanical energy.

All these effects are temperature- and time-dependent and are interrelated due to their common origin, but neither the elastomer nor the powderous filler alone shows such behaviour. The filler is composed of relatively stiff particles that do not undergo significant deformation by themselves. Also, the entropy-elastic behaviour of the elastomer or rubber matrix is quite well understood [3–6].

The physical origin of the stress-strain behaviour of filled elastomers has been identified in a dynamic creation and damage of filler-filler bonds, i.e. glassy polymer bridges of some nanometers in thickness [5, 7–13]. The thermodynamically irreversible process leads to a dissipation of mechanical energy, which is analogous to dissipative friction forces and is, therefore, called "internal friction". The damage of bonds is structurally reversible, which means that re-agglomeration of filler particles takes place, but the process requires either mechanical or thermal activation energies. In this context it is important to note, that the structure of the filler within the matrix is responsible for the mechanical reinforcement and stress-strain relation [14–24]. A steady-state stress-strain cycle results, when the same amount of bonds is damaged and recovered during a cycle [25].

If the filler volume fraction in a rubber compound goes beyond a critical value, a percolated filler network is formed during the vulcanization process. But, already under low deformations this network starts to break down, and the storage modulus drops from relatively high to low values. The remaining fragments are clusters of filler particles which are still capable of reinforcing the rubber matrix. The principal mechanism is called hydrodynamic reinforcement. The branched clusters which are relatively stiff immobilize a certain amount of rubber. So, the deformation is concentrated in the remaining part of the matrix, the local strain is higher than the global external strain, and the measured stress is accordingly increased.

With increased loading, clusters are successively being broken, and more rubber takes part in the deformation, which in turn leads to the observed stress softening. When the load is decreased, damaged clusters can re-agglomerate. The cyclic breakage and re-agglomeration of damaged, fragile clusters causes the observed hysteresis and the increasing loss angle with increasing amplitude of deformation.

Even if a rubber matrix shows little temperature and time-dependence above the glass transition region, the mechanical response of a filled elastomer always has a marked dependence on temperature. Also, the relaxation of stress or strain, after the state of the material has been brought out of equilibrium, shows a wide relaxation time spectrum. It seems that this behaviour is caused by the filler-filler bonds which are in a glassy like transient state.

Various models are found in the literature to describe the stress-strain-behaviour of unfilled as well as filled elastomers. It can be said that non of them fulfils the needs of the engineer who wants to have a tool to calculate the mechanical behaviour of a loaded component and its dependence on temperature, frequency and the history of deformation. Though the mechanical response of filled elastomers is not fully understood, it has been shown that the nanostructure of the filler particles and their interaction with the polymer are of basic significance. This in particular leads to a wide variation of material properties. In order to relate material parameters of rubber and filler to those of the vulcanized material we use a microstructure-based approach where material parameters are physical quantities, instead of mere fit parameters.

In the present report, at first, we will formulate the basic assumptions of the employed theory of reinforcement. Then, we will adapt the model equations to measured stress-strain cycles of unfilled and various filled elastomers, whereby compression and tension tests will be considered. In particular, we will focus our attention on temperature effects regarding stress softening, hysteresis and the inelastic setting behaviour of filled rubbers. The main part of the model which describes hydrodynamic reinforcement and stress softening has also been implemented into the Finite-Element Method (FEM). As one example, a rolling rubber wheel under load will be simulated. Results will be discussed in the frame of the physically well understood material parameters obtained from the fitting procedures.

2 Experimental

2.1 Sample Preparation

To identify parameter for the micromechanical model and to verify the model in further simulations, measurements were carried out with unfilled and filled rubber compounds. Carbon black (CB) and silica together with the coupling agent silane (Si 69) were mixed into different rubbers, i.e. butadiene rubber (BR, Buna CB 10), solution styrene butadiene rubber (S-SBR, VSL 5025-0) and ethylene-propylene-diene rubber (EPDM, Keltan 512), respectively. The corresponding mass fractions in phr (per hundred rubber) are listed in Table 1 which also includes the vulcanization agents, i.e. ZnO, stearic acid and a semi-efficient cross-linking system (sulfur + accelerator CBS), and anti-ageing IPPD. The compounding was prepared using an internal mixer, Werner & Pfleiderer GK 1.5E at 50 rpm.

The whole compound was rolled for a homogeneous distribution and then allowed to rest for 2 h, before the vulcanization time t_{90} at 160°C was determined with a Monsanto vulcameter. For shear testing, 2-mm plates were vulcanized in a heating press. Two circular discs, each 20 mm in diameter, were punched out and glued onto circular plates of a double-sandwich testing apparatus. For uniaxial testing, axial-symmetrical dumbbells were used.

Table 1 Components of the investigated compounds in phr.

sample	S-SBR 2525-0	S-SBR 5025-0	BR CB 10	EPDM K. 512	N115	N339	Silica	ZnO	stear. acid	Sulfur	CBS	IPPD
SBR	100							4	1	1.7	2.5	1.5
S60N1		100			60			4	1	1.7	2.5	1.5
S60N3		100				60		4	1	1.7	2.5	1.5
SB6R2		85	15			60		3	1	1.7	2.5	1.5
E6K				100			60	3	1	1.7	2.5	1.5

2.2 Multihysteresis Measurements

Uniaxial multi-hysteresis tests at $\dot{\varepsilon} \approx 0.01/\text{s}$ were carried out on dumbbells, 15 mm in thickness, using a Zwick 1445 universal testing machine. For strain measurement, two reflection marks were placed in a 15 mm distance. Multi-hysteresis means: at constant velocity up and down-cycles between certain minimum and maximum strains, $\varepsilon_{\min}$ and $\varepsilon_{\max}$, are carried out. This is done five times each step, and after every of such steps the boundaries of deformation are successively raised ($\varepsilon_{\max}$) or lowered ($\varepsilon_{\min}$), respectively. A typical set of steps would be: 1, 2, 5, 10, 20, 30, 40, 60, 80, 100% for $\varepsilon_{\max}$, while $\varepsilon_{\min}$ would be kept constant at 0% for purely tension mode testing. Only every 5th up and down-cycle are evaluated, which can be regarded as being steady-state in a good approximation.

Engineering (nominal) stress is measured quite easily by dividing the measured force by the initial cross-section area. But, to determine the nominal strain in the specimen is not as trivial. Our preferred method is to measure the displacement between two reflection marks sticked onto the surface and divide the value by the initial distance (≈ 15 mm). This works very well for flat specimens in uniaxial tension mode but not necessarily for dumbbells and in compression mode. The essential area of the employed dumbbell specimen is characterized by a length of parallelity, $l_0 = 21$ mm. Within this area there should be a volume of nearly constant deformation – only at high negative strains a barrel-like shape with in-homogeneous deformation is formed.

Optical detectors follow the two reflection marks within the homogeneous area as they move along the axis of measurement. It is necessary that the two light beams do not cross over each other, which is why there has to remain a certain distance between the marks, also during a compression test, limiting the strain measurement to a minimum of $\varepsilon \approx -20\%$.

A way out of this limitation is the calculation of ε from the cross-head displacement, Δy, divided by l_0. Unfortunately, a comparison with the optical method shows large differences, indicating that the real length of deformation is larger than the geometrically parallel length l_0. Also, the attempt to choose an appropriate l_0 that matches both methods was not successful under tension deformation, because the necessary l_0 varies with the current load. The reason being, that the specimens geometry changes non-linearly while being stretched, parts of the specimen are pulled out of the clamps, and friction forces cause additional hysteresis. Only in pure compression mode a calibration of the real strain by measuring only the cross-head displacement is feasible. The calibration factor k can then be calculated by relating the "optical strain" ε_{opt} to the "cross-head strain" $\varepsilon_{\text{trav}}$ in the common range between $\varepsilon = -20$ and 0%.

Measurements at varied temperature were carried out at 22, 50 and 80°C and with higher cross-head velocity ($\dot{\varepsilon} \approx 0.025/\text{s}$). Measurements at higher temperatures are problematic, because heating up of the heating room and the specimen takes 15–30 min, while the measurement itself takes even longer.

As a consequence, consciously only compounds with a relatively "flat" vulcameter curve at longer curing times have been investigated, to minimize a change of the cross-linking structure at higher temperatures and consequent influence on stress.

3 Theory

The Helmholtz free energy density of a filled elastomer constitutes the elastic material function and can be described as the sum of matrix and filler contributions [7, 10]:

$$W = (1 - \Phi_{\text{eff}})W_R + \Phi_{\text{eff}}W_A, \tag{1}$$

where W_R is calculated for the entropy-elastic rubber phase, and W_A is the energy-elastic contribution of the soft agglomerates (damaged clusters) of filler particles. Φ_{eff} stands for the mechanical effective filler volume fraction (which is larger than the real volume fraction, mainly because of rubber occluded in the aggregates). For an isothermal reversible process W equals the potential mechanical energy. Therefore, the respective stress components of rubber and filler can be obtained as the derivatives of W_R and W_A with respect to the deformation.

The proposed mechanical model is in accordance with the second law of thermodynamics which states that the entropy production of a closed system can only be positive. If there is no heat conduction, the entropy production density, multiplied by the absolute temperature, equals the dissipation function (the mechanical work irreversibly converted to heat). During a closed hysteresis cycle, this is the closed integral of $\sigma d\varepsilon$. For a thermodynamically reversible elastic material law, as employed for the rubber matrix, the integral of the up cycle equals the negative of the down cycle, which means no production of entropy. As we will see later, the stress contribution of the filler is always positive in the up-cycle and negative in the down cycle, which means that the spent mechanical work is always positive, and so is the entropy production.

To describe the hyperelastic behavior of the rubber matrix, we use the non-affine tube model with non-Gaussian extension [3–6] which has the following form:

$$W_R = \frac{G_C}{2}\left\{\frac{\left(\sum_{\mu=1}^{3}\lambda_\mu^2 - 3\right)\left(1 - \frac{T_e}{n_e}\right)}{1 - \frac{T_e}{n_e}\left(\sum_{\mu=1}^{3}\lambda_\mu^2 - 3\right)} + \ln\left(1 - \frac{T_e}{n_e}\left(\sum_{\mu=1}^{3}\lambda_\mu^2 - 3\right)\right)\right\}$$

$$+ 2G_e\left(\sum_{\mu+1}^{3}\lambda_\mu^{-1} - 3\right). \tag{2}$$

It takes into consideration the finite length of polymer network chains that leads to a statistical distribution of conformations of a chain that is not anymore Gaussian with respect to the chain's end-to-end distance. It has been shown to describe the rubber elasticity very well and is thoroughly physically motivated. So, we have three free parameters which are related to the rubber network chain structure:

- G_c, the cross-link modulus, proportional to the density of network junctions,
- G_e, the modulus of topological constraints, proportional to the entanglement density,
- $n_e/T_e = n$, number of chain segments between two successive trapped entanglements (n_e is the number of segments between entanglements and T_e is the trapping factor).

Due to the presence of the filler, only parts of the material volume are deformed under an external strain ε or relative deformation $\lambda = 1 + \varepsilon$. We describe the local strain as ε multiplied by a factor X.

When the filler network has been destroyed at low strains, we can assume the presence of filler clusters dispersed in the rubber matrix. A filler cluster is thought to be composed of spherical particles (with diameter d) and to have a diameter ξ_μ in each space direction μ. The relative size x_μ is the ratio ξ_μ/d. Describing the structure of a cluster by fractal geometry, the anisotropic elastic modulus $G_{A,\mu}$ of a cluster depends on its diameter in the respective direction [7, 17]:

$$G_{A,\mu} \cong \frac{\bar{G}}{d^3} \left(\frac{d}{\xi_\mu}\right)^{3+d_{f,B}}, \tag{3}$$

where $\bar{G}$ stands for an elastic constant and $d_{f,B}$ for the fractal dimension of the cluster backbone. Assuming cluster-cluster aggregation (CCA), $d_{f,B}$ amounts to ≈ 1.3 [14]. Also, a Smoluchowski type of cluster size distribution $\phi_\mu(x)$ in spatial direction μ, according to the kinetics of CCA [7], is adopted:

$$\phi_\mu(x) = \frac{4x}{\langle x_\mu \rangle^2} \exp\left[-2\frac{x}{\langle x_\mu \rangle}\right], \tag{4}$$

where $\langle x_\mu \rangle$ stands for the average cluster size. Previously, an isotropic cluster size distribution, $\langle x_1 \rangle = \langle x_2 \rangle = \langle x_3 \rangle \equiv x_0$, has been used. To account for a preconditioning of clusters under compression, we use an anisotropic cluster size distribution here, as derived by Witten et al. [18], which states that clusters deform like the specimen as a whole. This means: $\langle x_\mu \rangle = \lambda_{\mu,\min} x_0$, x_0 being the initial isotropic average cluster size and $\lambda_{\mu,\min}$ the minimal relative deformation that has occurred during the deformation history in space direction μ.

A crucial point in our approach which we call "dynamic flocculation model" are the bonds between filler particles. Under the load transferred from the matrix to the clusters the filler-filler bonds can break. When the load is decreased, damaged clusters can re-agglomerate, but the strength of the filler-filler bonds is reduced, compared to the virgin bonds of undamaged clusters.

Also, damaged clusters are softer and more elastically deformable. Cyclic breakdown (stress release) and re-agglomeration of soft clusters causes hysteresis. The following paragraph describes the stress contribution of the filler clusters. We apply this to isochoric uniaxial loading in 1-direction which fulfils the symmetry conditions $\lambda \equiv \lambda_1$ and $\lambda_2 = \lambda_3 = \lambda^{-1/2}$.

3.1 Stress Softening and Hysteresis

The stress-strain relation is formulated in main axes. Within this framework, the reversible fraction of the mechanical energy density spent at soft filler clusters has the following form:

$$W_A(\varepsilon) = \sum_{\mu}^{\partial \varepsilon_\mu/\partial t > 0} \frac{1}{2} \int_{x_\mu(\varepsilon_{\mu,\max})}^{x_\mu(\varepsilon)} G_A(x)\varepsilon_{A,\mu}^2(x,\varepsilon_\mu)\phi_\mu(x)dx, \qquad (5)$$

Here, we integrate over the re-agglomerated section of the cluster size distribution $\phi_\mu(x)$, in each space direction μ where clusters are being stretched (up cycle). In directions of negative strain rate we assume that there is no stress acting on soft clusters, because all of these have been broken in the preceding cycle and are now re-agglomerating. Additionally, we assume that clusters deform plastically when compressed in a certain direction. Therefore, to calculate the strain energy in the up cycle of a uniaxial test ($\partial \varepsilon_1/\partial t > 0$), we only consider the axial elastic cluster strains $\varepsilon_{A,1}$. To describe the down cycle ($\partial \varepsilon_2/\partial t, \partial \varepsilon_3/\partial t > 0$) the same isochoric state of the material shall be modeled by an equibiaxial compression test that starts at the end of the tension cycle, where still $\varepsilon_{A,2} = 0$.

The strain of the filler clusters is determined by a stress equilibrium between the rubber matrix and the clusters. For simplicity, we assume linear-elastic behaviour of the clusters:

$$G_A(x_\mu)\,\varepsilon_{A,\mu}(x_\mu,\varepsilon_\mu) = \hat{\sigma}_{R,\mu}(\varepsilon). \qquad (6)$$

Only the change of stress relative to the start of a cycle can cause an elastic cluster deformation. Because in every space direction we have a balance of stress between matrix and cluster,

$$\hat{\sigma}_{R,\mu}(\varepsilon) := \sigma_{R,\mu}(\varepsilon) - \sigma_{R,\mu}\left(\frac{\partial \varepsilon_\mu}{\partial t} = 0\right), \qquad (7)$$

which is the matrix stress relative to the start of the cycle where $\partial \varepsilon_\mu/\partial t = 0$. The elastic deformation of clusters is caused by this very "relative stress" which in turn depends on the hydrodynamic amplified local relative deformation $\kappa_\mu(\varepsilon)$. The relative stress is always positive, because only space directions with increasing strain are taken into account, and therefore the difference of stress to the start of the cycle > 0. Now, we calculate the cluster's stress in spatial direction ν:

$$\sigma_{A,\nu} = \frac{\partial W_A}{\partial \varepsilon_{A,\nu}} = \sum_\mu \frac{\partial W_A}{\partial \varepsilon_{A,\mu}} \frac{d\varepsilon_{A,\mu}}{d\varepsilon_{A,\nu}}$$

$$= \sum_\mu^{\partial \varepsilon_\mu / \partial t > 0} \hat{\sigma}_{R,\mu}(\varepsilon) \int_{x_\mu(\varepsilon_{\mu,\max})}^{x_\mu(\varepsilon)} \frac{d\varepsilon_{A,\mu}(x)}{d\varepsilon_{A,\nu}(x)} \phi_\mu(x) dx. \tag{8}$$

The matrix stress perpendicular to the loading direction in the uniaxial case, $\sigma_{R,2} = \sigma_{R,3}$, is calculated as the stress that is needed to cause the same deformation, from equivalence of energies. This gives:

$$\sigma_{R,2}(\varepsilon) = -\kappa_1(\varepsilon)^{3/2} \sigma_{R,1}(\varepsilon). \tag{9}$$

The integral in Equation (8) depends on the mutual derivative of cluster strains, $d\varepsilon_{A,\mu}/d\varepsilon_{A,\nu}$, and has to be evaluated in order to calculate the down cycle. Because the strain of the clusters $\varepsilon_{A,\mu}$ depends on the cluster size, we have no exact solution of this integral. Therefore, we use an approximation by replacing the derivatives of the cluster strains in the integral in Equation (8) by an average over the whole cluster ensemble. Assuming the re-agglomerated clusters, on average, deform like the specimen as a whole, this yields in the uniaxial case:

$$\left\langle \frac{d\varepsilon_{A,2}}{d\varepsilon_{A,1}} \right\rangle \approx \frac{d\varepsilon_2}{d\varepsilon_1} = -\frac{1}{2}(1+\varepsilon_1)^{-3/2}, \tag{10}$$

With this assumption the integral in Equation (8) can be solved analytically, if the Smoluchowski type cluster size distribution $\phi_\mu(x)$, Equation (4), is used. In the case of an isotropic cluster size distribution and uniaxial loading with $\lambda \equiv \lambda_1$ and $\lambda_2 = \lambda_3 = \lambda^{-1/2}$, this yields for the up cycle:

$$\sigma_{A,1}^{\text{up}}(\varepsilon) = \hat{\sigma}_{R,1}(\varepsilon) \left[\left(2\frac{x_1(\varepsilon_{\max})}{x_0} + 1 \right) \exp\left[-2\frac{x_1(\varepsilon_{\max})}{x_0} \right] \right.$$
$$\left. - \left(2\frac{x_1(\varepsilon)}{x_0} + 1 \right) \exp\left[-2\frac{x_1(\varepsilon)}{x_0} \right] \right] \tag{11}$$

and for the down cycle:

$$\sigma_{A,1}^{\text{down}}(\varepsilon) = -\frac{\hat{\sigma}_{R,2}(\varepsilon)}{(1+\varepsilon)^{3/2}} \left[\left(2\frac{x_2(\varepsilon_{\min})}{x_0} + 1 \right) \exp\left[-2\frac{x_2(\varepsilon_{\min})}{x_0} \right] \right.$$
$$\left. - \left(2\frac{x_2(\varepsilon)}{x_0} + 1 \right) \exp\left[-2\frac{x_2(\varepsilon)}{x_0} \right] \right]. \tag{12}$$

Because the relative stress is always positive, it turns out that the clusters give a positive stress contribution in the up cycle and a negative one in the down cycle.

The local relative deformation $\kappa_1(\varepsilon)$ is calculated from the global deformation and the strain amplification factor X. By this way, $\sigma_{R,\mu}$ and the cluster deformation depend on X which we calculate in the following section.

3.2 Hydrodynamic Strain Amplification

Huber and Vilgis calculated an amplification factor X for overlapping CB aggregates [19] assuming a fractal geometry. They found that X is proportional to powers of filler volume fraction and relative aggregate diameter. Because undamaged clusters are also stiff and show a fractal geometry, we can use the same result. All of the stiff clusters contribute to X, according to their diameter. Damage of stiff clusters causes stress softening by decreasing X, which is expressed as an integral over the "surviving", hard, section of the cluster size distribution, while broken clusters are treated as having a relative size of $x = 1$. Essentially, the strain amplification factor is a function of the maximum deformation the material has been subjected to during its entire deformation history [10]:

$$X_{\max} = X(\varepsilon_{\min}, \varepsilon_{\max})$$

$$= 1 + \frac{c}{3}\Phi_{\text{eff}}^{2/(3-d_f)} \sum_{\mu=1}^{3} \left(\int_0^{x_{\mu,\min}} x^{d_w-d_f} \phi(x)dx + \int_{x_{\mu,\min}}^{\infty} \phi(x)dx \right) \quad (13)$$

where the constant c is taken to be ≈ 2.5, the Einstein coefficient for spherical inclusions. The exponent d_w stands for the anomalous diffusion exponent which amounts to ≈ 3.1 [14], and d_f is the fractal dimension of the filler clusters (≈ 1.8 for cluster-cluster aggregation [15]).

To solve the integrals analytically, we make use of an approximation for the exponent: $d_w - d_f = 1$. This results in:

$$X_{\max} = 1 + \frac{c}{3}\Phi_{\text{eff}}^{2/(3-d_f)} \sum_{\mu=1}^{3} \left(1 + \int_0^{x_{\mu,\min}} (x-1)\phi(x)dx \right). \quad (14)$$

The solution of this expression in the case of uniaxial loading with $\lambda \equiv \lambda_1$ and $\lambda_2 = \lambda_3 = \lambda^{-1/2}$ is given as follows:

$$X_{\max} = 1 + \frac{c}{3}\Phi_{\text{eff}}^{2/(3-d_f)} \left\{ 3x_0 - \exp\left[-2\frac{x_{1,\min}}{x_0}\right] \right.$$

$$\left. \times \left[(x_0 - 1)\left(1 + \frac{2x_{1,\min}}{x_0}\right) + \frac{2x_{1,\min}^2}{x_0} \right] \right.$$

$$- 2\exp\left[-2\frac{x_{2,\min}}{x_0}\right]\left[(x_0-1)\left(1+\frac{2x_{2,\min}}{x_0}\right)+\frac{2x_{2,\min}^2}{x_0}\right]\right\}. \tag{15}$$

For an explicit evaluation of this strain amplification factor, we also have to calculate the integration limits $x_{\mu,\min}$. The tensile strength of damaged bonds s_d (which governs the amount of hysteresis) can be expressed by their failure strain $\varepsilon_{d,b}$ and elastic modulus Q_d/d^3, the same is valid for virgin bonds where we use the index "v". As depicted in [7], the cluster strain under a certain load rises stronger with cluster size than the failure strain does. Accordingly, with rising load, large clusters break fist followed by smaller ones. The critical size of currently breaking clusters was accordingly derived:

$$x_\mu(\varepsilon) = \frac{Q_d \varepsilon_{d,b}}{d^3 \hat{\sigma}_{R,\mu}(\varepsilon)} \equiv \frac{s_d}{\hat{\sigma}_{R,\mu}(\varepsilon)}. \tag{16}$$

The strength of virgin bonds s_v governs the strain amplification factor X, because it enters into the minimum size of damaged clusters $x_{\mu,\min}$ and consequently into the upper integration limit of Equation (14) and the solution of the integral, Equation (15):

$$x_{\mu,\min} = \frac{Q_v \varepsilon_{v,b}}{d^3 \hat{\sigma}_{R,\mu}(\varepsilon_{\max})} \equiv \frac{s_v}{\hat{\sigma}_{R,\mu}(\varepsilon_{\max})}. \tag{17}$$

The two parameters s_d and s_v, i.e. the tensile strength of damaged and virgin filler-filler bonds, can be treated as fitting parameters.

3.3 Constance of Volume

For unfilled as well as for filled elastomers it has been found experimentally that the volume remains more or less constant during deformation. This makes it possible to derive a uniaxial stress from e.g. Equation (2) as a total derivative to axial strain and to calculate mutual derivatives like in Equation (8). On the other side, the condition of constant volume has to be taken into consideration for strain amplification, if large deformations occur.

Mathematically, inner constant volume means:

$$\kappa_1 \kappa_2 \kappa_3 = (1+X_1\varepsilon_1)(1+X_2\varepsilon_2)(1+X_3\varepsilon_3) = 1, \tag{18}$$

and outer constant volume means:

$$\lambda_1 \lambda_2 \lambda_3 = (1+\varepsilon_1)(1+\varepsilon_2)(1+\varepsilon_3) = 1. \tag{19}$$

To fulfill these conditions, for deformations $\varepsilon > 0$, the amplification factors have to vary with strain. But, to identify the X_i, we need a third equation which derives from the following considerations. If for instance $X_1 = 10, \kappa_1$, as

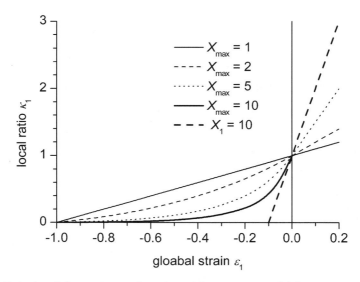

Fig. 2 Relative deformation within the rubber matrix, $\kappa_1(\varepsilon)$ for unixial deformation with varied $X_{\max}$, the thick dashed line indicates the local relative deformation if X_1 would be kept constant.

depicted in Figure 2, would take on negative values for $\varepsilon_1 < -10\%$, which is not permissible, physically. Therefore, the amplification factor for space directions with negative strain has to depend on strain. For simplicity, we assume that for the direction of largest strain: $X_f = $ const. (only dependent on maximum strain) and $X_f = X_{\max}$ is calculated from formulae (14) and (15). Using the assumption of constant volume and uniaxial or equibiaxial symmetry conditions, the other amplification factors can be calculated from that.

For the case of *uniaxial tension* in space direction 1 with the known symmetry conditions, and considering that $X_f = X_1 = X_{\max}$, the amplification factors in lateral directions 2 and 3 can be calculated from the equations above:

$$X_2 = X_3 = \frac{(1 + X_{\max}\varepsilon_1)^{-1/2} - 1}{(1 + \varepsilon_1)^{-1/2} - 1}. \tag{20}$$

This relation also holds for *equibiaxial compression* in directions 2 and 3. If the mode of loading is turned to *uniaxial compression* (or *equibiaxial tension*), we have $X_f = X_2(=X_3) = X_{\max}$, and the amplification factor in axial direction 1 is:

$$X_1 = \frac{(1 + X_{\max}\varepsilon_2)^{-2} - 1}{(1 + \varepsilon_2)^{-2} - 1} = \frac{(1 + X_{\max}((1 + \varepsilon_1)^{-1/2} - 1))^{-2} - 1}{\varepsilon_1}. \tag{21}$$

Whatever the value of X_2 now may be, X_1 will approach 1 when $\varepsilon_1 \to -1$, and accordingly $\kappa_1 \to +0$.

In order to unify the calculation for all required modes of loading, we use an alternative formalism by introducing an amplification exponent r, equal for all space directions:

$$\kappa_1 = \lambda_1^r, \quad \kappa_2 = \lambda_2^r, \quad \kappa_3 = \lambda_3^r.$$

$r = 1$ holds for unfilled rubbers. The formalism immediately satisfies inner volume constance, if outer volume constance is given. The amplification exponent, which is derived by elementary transformations, is dependent on $X_{\max}$ and current relative deformation in direction of maximum strain, λ_f:

$$\kappa_\mu = \lambda_\mu^r, \quad r = \frac{\ln(1 + X_{\max} \cdot (\lambda_f - 1))}{\ln(\lambda_f)}, \tag{22}$$

while, with the present formalism, for all modes of loading the same equations hold.

3.4 Dependence on Temperature

Already in the early days of the study of filled elastomers is has been noted that the elastic modulus shows an Arrhenius-activated temperature behaviour [1]. It is also established experimentally that reinforcement and hysteresis decrease with increasing temperature. This dependence on temperature neither stems from the filler particles by themselves nor from the rubber matrix (above the glass transition). We attribute the behaviour to a temperature-dependent elastic modulus Q/d^3 of the filler-filler bonds [10]. These bonds consist of glassy rubber due to the constrained mobility of polymer chains. This means that the deformation state is not in thermodynamic equilibrium but in a so-called "constrained equilibrium" where the internal state does not change significantly within a reasonable timescale. But, importantly, the relaxation to a "static" state is accelerated by thermal energy.

The storage modulus of virgin and damaged bonds can be described by an Arrhenius relation with respective activation energy. The bond strength can accordingly be computed. Assuming a constant failure strain of the bonds, we can write for the strength of *virgin* and *damaged* filler-filler bonds, respectively:

$$s_v \equiv \frac{Q_v \varepsilon_{b,v}}{d^3} = s_{v,\mathrm{ref}} \exp\left[\frac{E_v}{R}\left(\frac{1}{T} - \frac{1}{T_\mathrm{ref}}\right)\right],$$

$$s_d \equiv \frac{Q_d \varepsilon_{b,d}}{d^3} = s_{d,\mathrm{ref}} \exp\left[\frac{E_d}{R}\left(\frac{1}{T} - \frac{1}{T_\mathrm{ref}}\right)\right], \tag{23}$$

where E_v and E_d are the respective activation energies and T the absolute temperature. $s_{v,\mathrm{ref}}$ and $s_{d,\mathrm{ref}}$ are reference values of the bond's strength determined at $T = T_\mathrm{ref}$. Because the strength of virgin bonds s_v governs the strain amplification factor X, it is also responsible for stress softening and G'. In contrast, sd is responsible for hysteresis and G''.

In order to determine E_v and E_d, temperature-dependent vertical shift factors from viscoelastic master curves of $G'(f)$ and $G''(f)$ can be used. However, the thermal activation energies decrease with strain amplitude. This is why dynamic-mechanical tests limited in strain are not totally sufficient to determine values of E_v and E_d applicable for higher strains. Therefore, activation energies from vertical shift factors can only serve as a first approximation for a subsequent fitting procedure.

Filled elastomers also show a certain inelastic behaviour called setting which is characterized be a permanent deformation ε_{set} at released load. The corresponding stress contribution has to be described by a term separate from elastic potentials. The set stress σ_{set} is the stress required to get the material back into the un-deformed state. Because the physical mechanism of setting is not fully understood, we utilize a semi-empirical description. Empirically, a separate dependence of σ_{set} on maximum deformation and the distance of the temperature T from the glass transition temperature T_g is found:

$$\sigma_{set} = a(T - T_g) \cdot f(\varepsilon_{min}, \varepsilon_{max}). \tag{24}$$

The two functions a and f can be described by simple analytical expressions. This was verified for several CB-filled rubbers in tension and compression. In the whole expression there is only one free fitting parameter, $s_{set,0}$, which stands for the setting stress at $\varepsilon_{max} = 1$ (while $\varepsilon_{min} = 0$) and reference temperature (in our experiments, 22°C).

4 Results and Discussion

4.1 Uniaxial Compression-Tension Test of Unfilled Rubber

As can be seen from Figures 3 and 4, the non-affine tube model with non-Gaussian extension satisfactorily describes the stress-strain behaviour of the unfilled SBR. Because there is no filler present in the compound, up and down-cycle of the quasi-static measurement are almost equal. In the Mooney–Rivlin plot there is some scattering of data in the vicinity of $\lambda = 1$ zero strain) due to the normalization of stress to the Neo-Hooke term $\lambda - \lambda^{-2}$ which also becomes zero at $\lambda = 1$.

4.2 Adaptation of the Model for Various Filled Rubbers in Tension

Figure 5 demonstrates that a good adaptation for pre-conditioned samples can be obtained with a single set of polymer parameters ($G_c = 1.09$ MPa,

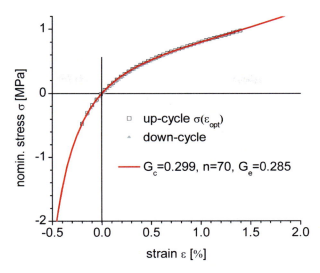

Fig. 3 Uniaxial stress-strain measurement and fit with tube model, $T = 22°C$, S-SBR.

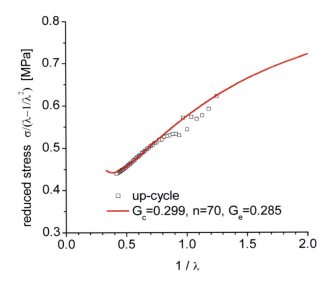

Fig. 4 Mooney–Rivlin plot of data above.

$G_e = 0.6$ MPa, $n_e/T_e = 45$), simply by varying the strain amplification factor X. The fitted parameters are $X_i = 7.520, 5.915, 4.733, 4.225$ and 3.694 for successive increasing pre-strain from $\varepsilon_{\max} = 10$ to 50%. The tube constraint modulus G_e was not treated as a fitting parameter but estimated from the plateau modulus $G_N \approx 1.2$ MPa ($G_e = G_N/2$). Further fitting parameters

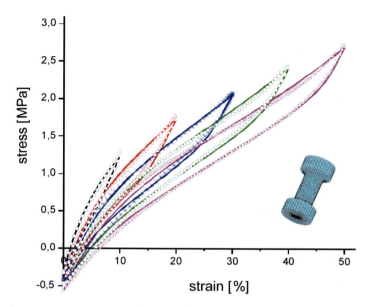

Fig. 5 Quasi-static stress-strain cycles in uniaxial tension (5th cycles) of E6K for pre-strains $\varepsilon_{\max}$ between 10 and 50% (symbols) and adaptations with the dynamic flocculation model (lines). Experimental data were obtained by using dumbbells as shown in the inset.

are the yield stress of damaged filler-filler bonds $s_d \equiv Q_d\,\varepsilon_{b,d}/d^3 = 31$ MPa, the effective filler volume fraction $\Phi_{\mathrm{eff}} = 0.26$ and the relative mean cluster size $x_0 \equiv \xi_0/d = 10.1$. Note that the value of Φ_{eff} is close to the filler volume fraction $\Phi \approx 0.22$, indicating that the reinforcing silica particles are almost spherical.

The simulation data shown in Figure 5 have been obtained without applying Equation (15), since the strain amplification factor X is treated as independent variable. However, it is found that the dependency of the fitting parameters X_i on $\varepsilon_{\max}$ is in fair agreement with the predicted behavior of Equation (15). This has been shown e.g. for carbon black and silica-filled EPDM and SBR rubbers [10]. The dependence equation (15) of the strain amplification factor $X_{\max} = X(\varepsilon_{\max})$ on pre-strain is depicted in Figure 6 for various filled elastomer materials, as indicated. It corresponds to the well known Payne effect of the storage modulus in the quasi-static limit, since $X_{\max}$ determines the initial slope of the hysteresis cycles, which can be compared with the storage modulus G' of non-linear viscoelastic materials. The data are obtained from adaptations to the 5th stress-strain cycles in the range between 1 and 100%, similar to the ones shown in Figure 5.

From the data in Figure 6 it becomes obvious that in the large strain regime $X_{\max}$ approaches a constant value scaling with the effective filler volume fraction Φ_{eff}, since the sum in Equation (14) approaches the value three in the

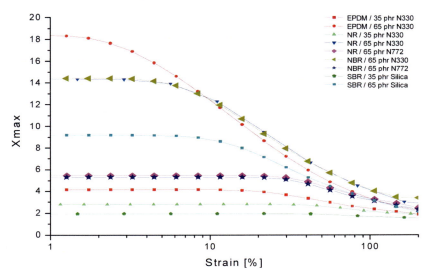

Fig. 6 Variation of the strain amplification factor X_{max} with pre-strain according to adaptations with Equation (15) to uniaxial stress-strain cycles at pre-strains between 1 and 100% of various filled rubber materials, as indicated.

limit of large stress values $\sigma_{R,1}$. For small pre-strains, the strain amplification factor levels out as well, since all clusters contribute to X_{max} in the limit of small stress values $\sigma_{R,1}$. Hence, the mean slope of the stress-strain cycles remains almost constant in the small strain plateau regime, which is typically observed for the Payne effect of reinforced rubbers. Dependent on the microstructure of the rubber and the type and amount of filler one observes characteristic differences that are also well known from dynamic-mechanical measurements of the Payne effect under harmonic excitations. Accordingly, the presented dynamic flocculation model provides a microstructure-based explanation of stress softening phenomena, also denoted as Mullins effect, as well as filler-induced hysteresis, which is found to be closely related to the Payne effect observed under dynamic loadings.

4.3 CB-Filled Rubber in Combined Compression-Tension Test

Multihysteresis measurements in combined compression-tension were carried out for the compound SB6R2, as described in Section 2.2 and depicted in Figure 1. Figure 7 shows the extracted 5th cycles and fits with the dynamic flocculation model, namely the summed-up contributions of matrix, cluster, and set -stresses, $\sigma_{R,1}+\sigma_{A,1}+\sigma_{set}$, vs. axial strain, as introduced in Section 3.

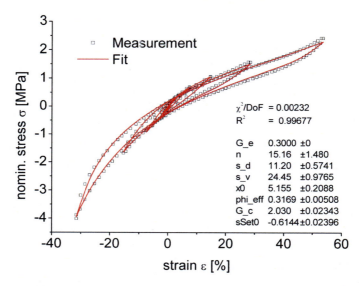

Fig. 7 Uniaxial multihysteresis measurement and fit, inset shows parameter set, $T = 22°\text{C}$, SB6R2.

The set of fitted parameters is given as an inset. Some parameters can be assumed to have specified values. These parameters are recognizable from the fitting error = 0 in the right column. The parameter $s_{\text{set},0}$ can be calculated from the evaluated set stresses and the glass transition temperature T_g of the rubber. For the factor c we assume the value $c = 2.5$ calculated by Einstein for delude rigid spheres. And the modulus of topological constraints, G_e, should be equal to $\approx 1/2\, G_N$, the plateau value of un-cross-linked rubber which is 0.6 MPa for SBR, but only under the assumption that there are no filler-induced entanglements.

Generally, the identified parameters are physically reasonable:

- $n_e/T_e \equiv n$ (number of chain segments between trapped entanglements) ≈ 15
- s_d and s_v amount to some 10 MPa, whereby: $s_d < s_v$
- $x_0 \approx 5$ is near a typical value of about 10 particles/cluster
- $\Phi_{\text{eff}} \approx 0.32$ is higher than the filler volume fraction, $\Phi = 0.24$, and a reasonable value for 60 phr of "structured" CB particles of grade N339
- G_c (cross-link modulus) ≈ 2 MPa is very high, compared to the unfilled rubber, an reflects effective cross-links induced by the filler surface

It has also been verified in previous studies that parameters fitted with the dynamic flocculation model vary systematically when the filler volume fraction is varied [27].

4.4 CB-Filled Rubber at Varied Particle Size and Temperature

For investigating the effect of particle size of CB, the compounds S60N1 and S60N3 were tested in multihysteresis tension mode at reference temperature $T_{\text{ref}} = 22°C$. The stress-strain curves (5th cycles) are depicted in Figures 8 and 9 (symbols), where fits with the dynamic flocculation model (lines) are also shown. The fitted parameters are listed in the legend of the plots, which again appear physically reasonable. For taking into account the effect of surface induced entanglements and for getting better fitting results, in this case the modulus of topological constraints, G_e, has also been treated as a fit parameter. It is found to converge at values of $G_e = 1.505$ MPa (S60N1) and $G_e = 1.257$ MPa (S60N3), respectively, which are larger than the previously used specified value $G_e = 0.3$ MPa of the polymer matrix. This indicates that the attractive interaction of the polymer chains with the filler surface increases the entanglement density significantly. We point out that this modification slightly affects the other parameters, but mostly the cross-link modulus G_c which is found to be reduced compared to the fits with $G_e = 0.3$ MPa.

The two compounds differ in the kind of CB used: N115 (S60N1) and N339 (S60N3). The different morphology of the filler has a considerable effect on the stress-strain behaviour and accordingly on the fit parameters: N115 consists of smaller filler particles with a higher specific surface area. This results in a higher relative cluster size x_0 and Φ_{eff}, at constant Φ,

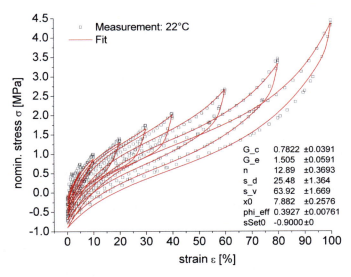

Fig. 8 Uniaxial multihysteresis measurement and fit, inset shows parameter set, S60N1.

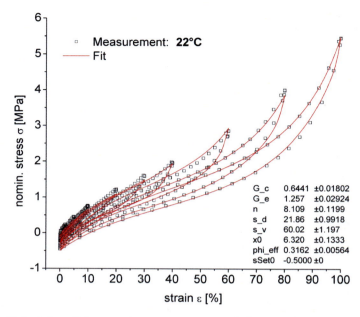

Fig. 9 Uniaxial multihysteresis measurement and fit at reference temperature, inset shows parameter set, S60N3.

compared to N339 (N115/N339: $x_0 = 7.882/6.32$, $\Phi_{\text{eff}} = 0.3927/0.3162$). N115 is also a more "active" filler, which apparently leads to stronger filler-filler bonds (N115/N339: $s_d = 25.48/21.86$ MPa, $s_v = 63.92/60.02$ MPa). But, the filler also affects the parameters of the polymer network: the higher filler surface area of N115 results in higher cross-linking and entanglement-densities (N115/N339: $G_c = 1.505/1.257$ MPa, $G_e = 1.505/1.257$ MPa). Both of these effects can be related to the higher specific surface of N115. If more entanglements are present one would also expect a lower n (segments between trapped entanglements), but the trapping factor, T_e, may also be reduced leading to a higher n for N115: $n = 12.89$, compared to $n = 8.109$ for N339.

For the composite with the less active CB N339, tension tests were also carried out at elevated temperatures 50°C and 80°C. Based on the fitting parameters obtained at 22°C, the stress-strain curves were simulated and compared to the experimental data in Figures 10 and 11.

Experimentally, a clear tendency is seen at higher temperatures: stress, in particular maximum stress and absolute value of set stress, and hysteresis are decreasing (Figures 9 to 11). The temperature dependences of softening and hysteresis are described by the Arrhenius-activated strength of virgin and damaged filler-filler bonds, respectively, Equation (23). To simulate the stress-strain curves at 50°C and 80°C activation energies of $E_v = 3$ kJ/mol and $E_d = 7$ kJ/mol have been chosen to achieve a fair agreement between

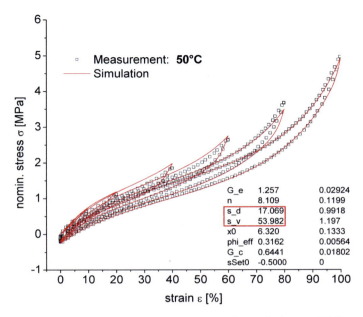

Fig. 10 Uniaxial multihysteresis measurement and simulation at 50°C, varied parameters marked, S60N3.

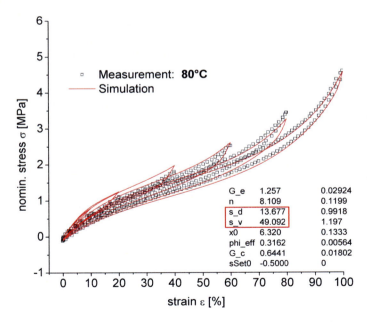

Fig. 11 Uniaxial multihysteresis measurement and simulation at 80°C, varied parameters marked, S60N3.

experiment and simulation. Using these activation energies, the strengths of filler-filler bonds, s_v and s_d, are calculated for the desired temperature T. The corresponding parameter values are marked in the insets of Figures 9 to 11.

However, it is obvious that higher thermal activation energies are necessary at lower strains. This corresponds to dynamic-mechanical analysis (DMA) measurements by which it is also possible to obtain activation energies. The procedure assumes the possibility to create viscoelastic master curves from a number of frequency sweeps at various temperatures. Using the time-temperature superposition principle quantified by the WLF equation, the single sweeps in the same frequency range (0.01–100 Hz) are aligned in a broader frequency spectrum by applying temperature-dependent "horizontal" shift factors. Because the elastic modulus of filled elastomers is not only dependent on the rubber matrix but additionally on the filler, vertical shift factors are introduced for the storage modulus G' and loss modulus G'', separately [11]. The vertical shift factors show the known Arrhenius dependency when the temperature is above the glass transition region of the matrix.

The crucial point is: G' is determined by stress softening and therefore has the same dependence on temperature as sv with the same E_v. The equivalent is valid for G'' and E_d, concerning hysteresis. For a S-SBR filled with 60 phr of N339 (S60N3) master curves were created from DMA at 3.5% strain amplitude [12]. The calculated thermal activation energies, $E_v = 5.4$ kJ/mol and $E_d = 7.72$ kJ/mol, are higher than the values necessary to simulate the large-strain behaviour (but sufficient for strains below $\varepsilon \approx 40\%$). The activation energies at 0.5% strain amplitude are, consequently, even higher: $E_v = 8.81$ kJ/mol and $E_d = 15.26$ kJ/mol.

Figure 12 shows the set stress, σ_{set}, for several temperatures as a function of maximum strain. The empirical fit function has the same form for every temperature, only scaled by a temperature-dependent factor $a(T - T_g)$, according to Equation (24). The normalized σ_{set} values, more or less, fall together onto a single master curve.

4.5 Finite-Element (FE) Simulation of a Rolling GROSCH Wheel

The main part of the flocculation model which describes hydrodynamic reinforcement and stress softening has been implemented into the Finite-Element Method (FEM). This was done by referring to a strain amplification exponent (damage parameter), which has been described previously in more detail [28]. Here, we demonstrate for the first time the evolution of this damage parameter under practical conditions, i.e. for a rolling rubber wheel (known as "Grosch wheel") whose dimensions are depicted in Figure 13a. This rubber wheel was meshed as shown in Figure 13b with symmetric boundary conditions in 3-direction.

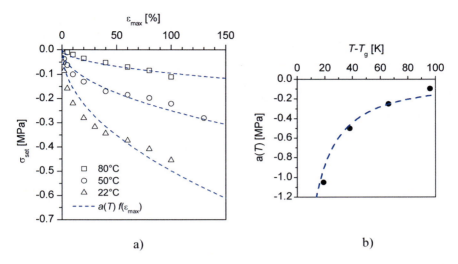

Fig. 12 (a) Set stresses for several temperatures vs. maximum strain, fits as dotted curves, (b) factor $a(T - T_g)$ vs. temperature difference $T - T_g$, T_g being the glass transition temperature, S60N3.

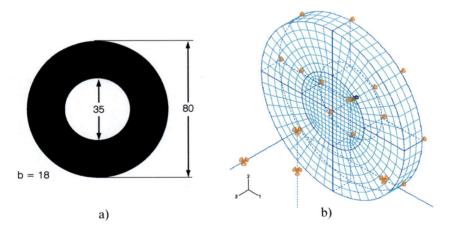

Fig. 13 (a) Geometry of a GROSCH wheel, (b) FE mesh.

At the beginning of time history, a compressive load of $F_{\max} = 1.9$ kN in 2-direction is applied to the middle part which consists of a rigid shaft. Under the load the wheel is compressed by 7.8 mm, as can be seen in Figure 14. In the following step the wheel rotates around the axis and rolls on the ground, frictionless. Despite the constant load, the displacement increases by 1.2 mm (between the first and last snapshot) while the wheel fulfils one complete revolution. This is due to stress softening, as can be seen from the strain

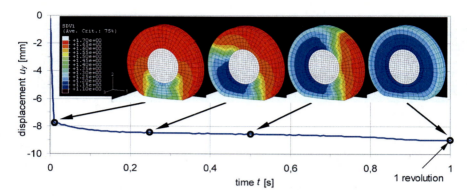

Fig. 14 Rigid displacement of the wheel as a function of time, r is depicted in the insets for chosen times.

amplification exponent, r. Prior to loading, r is ubiquitously equal to 1.7. It can be observed how it decreases during the test. The smallest values of 1.1 are found near the shaft where, simultaneously, the highest stresses are found. Because the decrease of r corresponds to stress softening, it can be regarded as kind of damage variable.

5 Conclusions

The dynamic flocculation model incorporates hydrodynamic reinforcement of the rubber matrix by a fraction of rigid filler clusters, which decreases with increasing strain leading to the characteristic stress softening effects of filler reinforced rubbers. Furthermore, the hysteresis is described by cyclic breakage and re-agglomeration of filler clusters. The temperature dependence of the stress-strain behaviour in the rubber elastic plateau area (well above the glass transition temperature) is dominated by that of the filler network. Accordingly, all temperature effects are described via an Arrhenius-like thermal activation of the stiffness and strength of filler-filler bonds. The (plastic) set behaviour is taken into account empirically by simple analytical functions depending on temperature and maximum deformation.

Using dumbbell specimens, we have performed uniaxial stress-strain measurements of variously filler rubber samples at room temperature $T_{\text{ref}} = 22°\text{C}$. Parameterfits show that the model satisfactorily describes tension tests up to 100% strain as well as combined compression-tension tests between -20 and 60% strain. Due to microstructure-based modeling, physical material parameters are obtained by parameter fitting of uniaxial tests. The parameters lie in a physically reasonable range. The filler surface evidently increases cross-link as well as entanglement-density of the SBR/BR rubber matrix. The different morphology of two carbon blacks, N115 and N339, is consistently reflected in the fit parameters. Also, tension tests at elevated temperatures, $T = 50°C$ and $T = 80°C$, were carried out. The experiments show that stress and hysteresis decrease at higher temperatures. The behaviour could be simulated from a parameter fit of measurements at T_{ref}, when the strength of virgin and damaged filler-filler bonds, respectively, is described by an Arrhenius relation and activation energies are determined. The thermal activation energies can be evaluated from dynamic mechanical data, where the temperature dependence of the filler network becomes apparent though they depend somewhat on strain energy and somewhat higher activation energies are necessary at lower strains.

Finally, the implementation of the model into a Finite Element algorithm has been addressed by referring to a strain amplification exponent as some kind of damage parameter. As an example, a rolling rubber wheel under constant load has been simulated with this FE algorithm, demonstrating the history of stress softening via the evolution of the strain amplification exponent. Dependent on the applied load, this damage parameter is found to decrease significantly in regions where large stresses are present.

Acknowledgements

This work has been supported by the Deutsche Forschungsgemeinschaft (DFG) as a part of the collaborate research unit DFG-Forschergruppe "Dynamische Konktaktprobleme mit Reibung bei Elastomeren" (FOR 492). This is gratefully acknowledged.

References

1. Kraus, G. (ed.): Reinforcement of Elastomers. Wiley, Interscience Publ., New York (1965)
2. Medalia, A.I.: Elastic modulus of vulcanizates as related to carbon black structure. Rubber Chem. Technol. 46, 877 (1973)
3. Heinrich, G., Straube, E., Helmis, G.: Rubber elasticity of polymer networks: Theories. Adv. Polym. Sci. 85, 33–87 (1988)
4. Heinrich, G., Kaliske, M.: Theoretical and numerical formulation of a molecular based constitutive tube-model of rubber elasticity. Comput. Theor. Polym. Sci. 7(3/4), 227–241 (1997)
5. Klüppel, M., Schramm, J.: A generalized tube model of rubber elasticity and stress softening of filler reinforced elastomer systems. Macromol. Theory Simul. 9, 742–754 (2000)
6. Klüppel, M., Menge, H., Schmidt, H., Schneider, H., Schuster, R.H.: Influence of preparation conditions on network parameters of sulfur-cured natural rubber. Macromolecules 34(23), 8107–8116 (2001)
7. Klüppel, M.: The role of disorder in filler reinforcement of elastomers on various length scales. Adv. Polym. Sci. 164, 1–86 (2003)
8. Klüppel, M.: Hyperelasticity and stress softening of filler reinforced polymer networks. In: Macromol. Symp., vol. 200, pp. 31–43 (2003)

9. Klüppel, M., Meier, J., Heinrich, G.: Impact of pre-strain on dynamic-mechanical properties of carbon black and silica filled rubbers. In: Busfield, Muhr (eds.) Constitutive Models for Rubber III, p. 333. Swets & Zeitlinger, Lisse (2003)
10. Klüppel, M., Meier, J., Dämgen, M.: Modelling of stress softening and filler induced hysteresis of elastomer materials. In: Austrell, Kari (eds.) Constitutive Models for Rubber IV, p. 171. Taylor & Francis, London (2005)
11. Le Gal, A., Klüppel, M.: Modeling of rubber friction: A quantitative description of the hysteresis and adhesion contribution. In: Austrell, Kari (eds.) Constitutive Models for Rubber IV, p. 509. Taylor & Francis, London (2005)
12. Le Gal, A., Yang, X., Klüppel, M.: Evaluation of sliding friction and contact mechanics of elastomers based on dynamic-mechanical analysis. J. Chem. Phys. 123, 014704 (2005)
13. Klüppel, M., Meier, J.: Modeling of soft matter viscoelasticity for FE-applications. In: Besdo, D., Schuster, R.H., Ihlemann, J. (eds.) Constitutive Models for Rubber II, p. 11. Balkema, Lisse (2001)
14. Klüppel, M., Schuster, R.H., Heinrich, G.: Structure and properties of reinforcing fractal filler networks in elastomers. Rubber Chem. Technol. 70, 243 (1997)
15. Meakin, P.: Fractal aggregates. Adv. Colloid Interface Sci. 28, 249 (1988)
16. Havlin, S., Bunde, A. (eds.): Fractals and Disordered Systems. Springer, Heidelberg (1991)
17. Kantor, Y., Webman, I.: Elastic properties of random percolation systems. Phys. Rev. Lett. 52(21), 1891–1894 (1984)
18. Witten, T.A., Rubinstein, M., Colby, R.H.: Reinforcement of rubber by fractal aggregates. J. Phys. II France 3, 367–383 (1993)
19. Huber, G., Vilgis, T.A.: Universal properties of filled rubbers: Mechanisms for reinforcement on different length scales. Kautsch. Gummi Kunstst. 52(2), 102 (1999)
20. Klüppel, M., Heinrich, G.: Fractal structures in carbon black reinforced rubbers. Rubber Chem. Technol. 68, 623–651 (1995)
21. Heinrich, G., Klüppel, M.: Recent advances in the theory of filler networking in elastomers. Adv. Polym. Sci. 160, 1–44 (2002)
22. Heinrich, G., Klüppel, M., Vilgis, T.A.: Reinforcement of elastomers, current opinion in solid state and materials. Science 6, 195–203 (2002)
23. Klüppel, M.: Elasticity of fractal filler networks in elastomers. Macromol. Symp. 194, 39–45 (2003)
24. Heinrich, G., Klüppel, M.: The role of polymer-filler interphase in reinforcement of elastomers. Kautsch. Gummi Kunstst. 57, 452–454 (2004)
25. Kraus, G.: Mechanical losses in carbon-black-filled rubbers. J. Appl. Polym. Sci., Appl. Polym. Symp. 39, 75–92 (1984)
26. Lion, A.: Phenomenolocial modelling of strain-induced strucural changes in filler-reinforced elastomers. In: Proceedings 6th Rubber Fall Colloquium, Hannover, p. 119 (2004)
27. Klüppel, M.: Evaluation of viscoelastic master curves of filled elastomers and applications to fracture mechanics. J. Phys. Condens. Matter 21, 035104 (2009)
28. Klauke, R., Meier, J., Klüppel, M.: FE-implementation of a constitutive model for stress softening based on hydrodynamic reinforcement. In: Boukamel, A., et al. (eds.) Constitutive Models for Rubber V. Taylor & Francis, London (2008)

Multi-scale Approach for Frictional Contact of Elastomers on Rough Rigid Surfaces

Jana Reinelt and Peter Wriggers

Abstract. Within the analysis of many frictional contact problems Coulomb's law with a constant friction coefficient μ is used. This assumption is sufficient for many applications in structural mechanics; however in the special case of rubber friction on rough surfaces the resulting simplification cannot be accepted. The physical interactions between e.g. a tire and a road surface are very complex and still widely unknown. The elastomer undergoes large deformations during contact, such that the frictional properties result for the main part from internal energy dissipation and not just from the combination of surfaces in contact like e.g. adhesion. As it is apparent from experiments, the friction coefficient depends heavily on various parameters like sliding velocity, surface roughness, normal forces and temperature change. In this contribution a multi-scale approach will be used to make an attempt to understand the sources of elastomere friction in a more rigorous way.

1 Introduction

Rubber friction phenomena appear in various forms of industrial applications, e.g. seals, band conveyors, squeaking windshield wipers or tires on road tracks. The analysis of all these problems has different goals, among them could be acoustic aspects, wear and abrasion or the derivation of friction laws. The aim of this work is the investigation of the frictional behavior of a rubber block (e.g. the tread profile of a tire) on a rough surface like a road track. The roughness of the surface has essential influence on the frictional resistance. Because of its relative softness the rubber undergoes large deformations and is therefore able to adapt to the asperities of the track. The degree of flexibility

---

Jana Reinelt · Peter Wriggers
Institut für Kontinuumsmechanik, Gottfried Wilhelm Leibniz Universität
Hannover, Appelstraße 11, 30167 Hannover, Germany
e-mail: `jana.reinelt@gmx.net, wriggers@ikm.uni-hannover.de`

of the material depends in turn on multiple factors. Due to the viscoelastic behavior of the rubber, its reaction is much stiffer in case of large sliding velocities than for small speeds. The actual temperature of the material plays an important role and changes mechanical properties; in turn the temperature will be influenced by the amount of heat which is generated due to friction. The possible influence of rainwater and lubrication films between road and rubber increases the complexity of the problem. Adhesional effects may have great relevance, wear and abrasion of the rubber changes its profile and during extreme brake applications even the track profile may be battered.

Practical experiments are often limited in their results. Experimental rigs have mostly crucial restrictions concerning the applicable sliding velocity or the normal pressure. Exact temperature measurements in the contact zone are always difficult or sometimes even impossible; the penetration depth and true contact area can only be estimated and the repetition of measurements will often take place under different environmental conditions. During the performance of a measurement it is in most cases impossible to separate the various effects which influence the frictional behavior. Local heating cannot be excluded, the rubber may suffer from microcracks or the surface characteristics change because of dust or lubrication. Practically none of these influences can be switched off or separated in real-life experiments. Thus the interpretation of experiments suffers from the complexity and inseparability of the multiple conditions related to material and environment. Hence it is necessary to understand the fundamental mechanisms of rubber friction by dividing the topic into different subareas to seal off single effects.

A numerical simulation using the Finite Element Method seems to be ideally suited to understand the basic principles of rubber friction. Numerical tests are repeatable under slightly different conditions, whereas the user is able to select the effects which are interesting. However, computational methods represent only virtual reality and the results depend on the model definitions. In this work we try to find adequate models and formulations to simulate the most important effects of rubber friction and illuminate connections between them. Beside the derivation of a friction law using numerical calculations, the relations between the different friction contributions and their relevance should become clearer.

The surface of a road track has fractal character, see [31]. It is rough on many different length scales and each of them contributes to rubber friction. Despite the rapidly growing computational power a reasonable finite element analysis of such a model which includes three to four different length scales is still impossible. A special multiscale method is required to determine the effects from micro-roughness of single asperities to macro-roughness of the street. As the interesting length scales range from some micrometers to a few millimeters, a singular scale transition is not sufficient, but several intermediate steps are necessary. Based on the roughness scale a particular frequency of the rubber is excited. Hence different roughnesses yield a wide-spread loading frequency range which is important due to the damping characteristics of the elastomer. It is well known that the hysteretic frictional resistance depends directly on the energy dissipation inside the material due to frequency loading, see e.g. [17].

Adhesion is often cited to be another important contribution to rubber friction, although the magnitude of this effect remains unanswered up to now. In this work the contact model is extended by an adhesional part, which offers the possibility to analyze this question with numerical methods. Another indisputable influence is the temperature change inside the rubber material which can lead to significant changes of the frictional resistance.

Early work and large experimental tests concerning rubber friction in particular were done by Schallamach [36] and Grosch [9] in the middle of the last century. The basic results of these experiments are still valid in current models. Newer friction theories are based on physical models for small deformations which consider the roughness characteristic of the surface and the viscoelastic material properties, see [17] or [31].

Material models for rubber cover a wide spectrum, which includes hyperelastic nonlinear models [27] but also viscoelastic and plastic effects (e.g. [38] or [34]). A molecular motivated tube-model for rubber was introduced by Edwards in 1988 [5] and advanced for filled elastomers by Heinrich and Kaliske [12] in 1998. The model is based on the network structure of the polymer chains, which are restricted in their movement because of tangling and linking points. Meier et al. explained the Mullins effect with braking and reaggregation of filler clusters [22]. Another nonlinear, inelastic model for the description of filled elastomers was proposed by Ihlemann in 2003 [13]. In contrast to other approaches this model is not based on a free energy function but directly composed out of varying stress parts.

Some of these models are able to describe special characteristics of rubber like Mullins [24] and Payne [29, 30] effect, others prioritize viscous or plastic properties. Based on the assumption that the internal energy dissipation is the main factor of rubber friction, a nonlinear viscoelastic approach is used in this work.

In a more general setting, this approach corresponds to contact homogenization where one attempts to formulate a macroscopic contact law based on microscale information such as the topography and constitutive properties of the contacting surfaces, see [39, 40]. Such formulations have been carried out in the context of asperity based small deformation contact by, among others, Tworzydlo et al. (1998) where macroscale frictional and normal contact behaviors of randomly rough surfaces were analyzed, by Haraldsson and Wriggers [10] to formulate the load and slip distance dependency of frictional behavior between concrete and soil, by Bandeira et al. [2] to construct a micromechanically motivated law to describe the variation of normal pressure during contact and recently by Orlik et al. [28] to simplify the analysis of replacement/bone contact. See also [4] for a multiscale analysis of a wear problem concerning wheel/rail interaction.

2 Multiscale Approach

The nature of the contact surfaces plays an important role in almost every frictional problem. Smooth surfaces are in general expected to generate less tangential stress compared to a rough surface, if adhesional effects are neglected. The asperities of rough surfaces are predestined to interlock into each other during sliding. This increases the frictional resistance immensely, because each asperity is able to transmit large stresses in the local normal, but global tangential direction. If the stress exceeds a critical value the interlocking will break and the bodies slide.

Even though the rubber surface under investigation is here assumed to be perfectly smooth, it adjusts to the road surface under the applied pressure and thus undergoes locally large strains. Furthermore the surface asperities induce a cyclic loading of the sliding block, which leads for the chosen material to large viscoelastic dissipation effects. Those are considered to be the main reason for the large observed friction values of rubber-track combinations.

The roughness of road tracks can vary strongly e.g. for asphalt and concrete pavements. Measured friction values on such surfaces differ as well. Hence the magnitude of roughness seems to be very important for the frictional behavior. Thus its proper description is necessary. In this work focus is placed on abrasive paper because experiments have been carried out on this surface which then can be used to validate the numerical simulations. Both, abrasive paper and realistic road tracks have similar fractal characteristics. This yields a self-affinity over multiple scales, i.e. a magnification of a surface segment has similar roughness characteristics than the original surface. For mathematical fractals like the Mandelbrot set [21] or the Koch snowflake the refinement proceeds infinitely to smaller scales. Examples for natural fractals are the coast line or a road surface, whose self-affinity ends at the atomic scale. The influencing magnitudes of the different scales on friction are a priori unknown. All asperities excite the sliding rubber sample on varying frequencies and influence the frictional behavior. Therefore a detailed description of the surface geometry on all scales is essential. Persson stated a lower limit of influence for the road track wavelengths to be in the micrometer range [31]. At lower scales dust and dirt particles which fill the smaller cavities of the track and as a result roughness below this scale can be excluded. The upper cut-off length is usually in the magnitude of some millimeters; larger wavelengths may influence the whole tire behavior up to the wheel suspension, but have no consequences on the local friction behavior.

Within the range of the lower and upper limits for the fractal the surface pattern has a detailed form which has to be discretized for a finite element calculation. In order to pre-estimate the magnitude of the system of equations which has to be solved during a complete simulation a resolution of 25 μm is assumed. This leads to a pattern with 160,000 different height points on a square centimeter. The finite element model of a rubber block which is pressed on such a surface has to contain a multiple of degrees of freedom to reproduce the correct deformations of the material. Although the computational resources have been increasing rapidly during the last years, it is still not reasonable to solve such a problem with one FE model.

Hence it is advantageous to use a multiscale approach for the prediction of the frictional behavior using finite element simulations. Fractal surfaces can easily be conveyed into a set of superposed harmonic functions by using a Fourier-Transformation. To keep the calculations in a manageable framework, the number of approximation functions and consequently the number of scale transition steps was restricted. A small number of functions were supposed to cover the whole spectrum of length scales which may contribute to rubber friction. However, the different scales cannot be viewed independently from each other. The whole problem is highly nonlinear both due to contact conditions and large deformations, and simple superposition techniques are not acceptable. Furthermore the single scales are not independent, neither in amplitude nor in frequency. Due to these conditions, a simple addition of the friction contributions on the different scales is not sufficient; instead a multiscale model is presented in which the results of each scale are transmitted to the next larger scale in form of a local friction law, whereas the smaller roughness is smoothed in return. Within this approach the energy dissipation in the elastomere due to the excitation related to small wavelength at smaller scales is introduced. This procedure allows computations on relative smooth and regular surfaces with reproducible results.

Next the multiscale scheme is described which is followed by a method to specify the roughness characteristic of a track and the crossover and simplification to a smoother reproduction. The numerical test set-up is then formulated and the method is applied to an abrasive paper surface of corundum 400 using two different wavelengths as lower limits.

2.1 Formulation of the Multi-scale Approach

As described before, a sequence of computations at different scales has to be performed in order to obtain the total friction law at the macro-scale for elastomere friction. In this framework the local friction law on each scale is deduced by the results computed at the respective lower scale. Analogous to the concept of the Representative Volume Elements (RVE) known from homogenization methods for solids, see e.g. [45]), a so-called Representative Contact Element (RCE) is introduced to obtain homogenized results for the frictional resistance at each scale. In the following the RCEs are numbered corresponding to the current scale represented by the surface function z_i which will be described in Section 4.

The method starts at the smallest scale where the friction law is determined just from the bulk properties of the elastomer sample, see left-hand

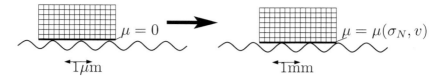

Fig. 1 Idea of scale transition.

side of Figure 1. Since frictionless contact is considered on that scale the energy dissipation results purely from the viscoelastic damping properties of the rubber. At this scale different numerical simulations are then carried out for different sliding velocities in horizontal direction and various vertical pressures. By this procedure the dependency of the resulting homogenized friction law on the sliding velocity and pressure can be deduced.

The finite element computation is performed by solving the weak form of the equilibrium including the contact part. The latter is on this smallest scale restricted to the frictionless normal contact formulation.[1] The weak form with respect to the initial configuration follows with the index $i = 1$ for the lowest scale as

$$\int_{\mathcal{B}_1} (\mathbf{S} : \delta \mathbf{E} - \rho_0 \bar{\mathbf{b}} \cdot \delta \mathbf{u}) \, dV - \int_{\Gamma_{t1}} \bar{\mathbf{t}} \cdot \delta \mathbf{u} d\Gamma + \int_{\Gamma_{c1}} p_N \delta g_N d\Gamma = 0. \quad (1)$$

Here $\mathbf{S}$ is the second Piola–Kirchhoff stress, $\mathbf{E}$ denotes the Green strains, $\rho_0 \bar{\mathbf{b}}$ the body forces and $\bar{\mathbf{t}}$ the applied surface tractions. $\delta \mathbf{u}$ are the test functions and p_N is the normal contact presure. While the block discretized by finite elements is pulled over the surface, the resulting forces on the upper side of the block can be computed. The sum of the vertical forces naturally matches the applied pressure multiplied by the top area of the RCE

$$\sum F_V = \sigma_V \, A^{\text{top}} = \langle p_N \rangle A^{\text{top}} \quad \text{with} \quad \langle p_N \rangle = \frac{1}{\Gamma_{c1}} \int_{\Gamma_{c1}} p_N d\Gamma. \quad (2)$$

These forces are equivalent to the normal forces at the contact area and thus represent the total contact force of the RCE_1.

The horizontal forces arise exclusively from the viscoelastic damping of the elastomer. Numerical tests with elastic material models did not lead to any horizontal resistance. A geometric influence due to shear deformation and piling up of the rubber on the slopes is insignificant here, because of the frictionless contact. The progress of the horizontal reaction forces is similar in all simulations. After a short phase of pure sticking, where the forces rise linearly and a transition period, the system reaches a steady state. The sum of the horizontal forces is averaged over the sliding distance related to one wavelength

---

[1] When adhesional effects are added in Section 7, the local friction on the smallest scale results from adhesive forces.

$$\left\langle \sum F_H \right\rangle = \frac{1}{l_s}\left(\sum F_H\right), \qquad (3)$$

where l_s denotes the sliding distance.

The resulting friction coefficient at the smallest scale yields

$$\langle \mu \rangle_i = \frac{\langle \sum F_H \rangle}{\sum F_V}, \qquad (4)$$

which depends on the sliding velocity v and the averaged contact pressure $\langle p_N \rangle$. Now an analytical friction function $\langle \mu, \rangle(\langle p_N \rangle, v)$ is fitted to the simulation results and transmitted to the next larger scale. The microasperities are smoothed and as a substitute the derived friction law is applied locally at each contact element, see right-hand side of Figure 1. The local contact pressure is denoted by p_N to mark the difference to the averaged contact stress $\langle p_N \rangle$ which is predetermined, see (2), and constant in each simulation. The local pressure p_N instead varies in each contact element and is calculated during the FE analysis. The same is true for the velocities. The form of the analytic fitting function for the frictional contact law depends on the used material and the surface characteristics. In this context we restrict to a general form for the friction coefficient in the contact element[2]

$$\mu_i^{\text{loc}} = \langle \mu \rangle_{i-1}(p_{N_i}, v_i^{\text{loc}}). \qquad (5)$$

Thus the friction coefficient of the local friction law at level $i+1$ which stems from the homogenization (4) depends on the local pressure and sliding velocity at $i+1$. On the larger scales $i \geq 2$ the algorithm changes in terms of the contact formulation. The tangential contact part due to friction has now to be included into the weak form

$$\int_{\mathcal{B}_i}(\mathbf{S}:\delta\mathbf{E} - \rho_0\mathbf{b}\cdot\delta\mathbf{u})\,\mathrm{d}V - \int_{\Gamma_{ti}}\bar{\mathbf{t}}\cdot\delta\mathbf{u}\,\mathrm{d}\Gamma + \int_{\Gamma_{ci}}(p_N\delta g_N + \mathbf{t}_{T_i}\cdot\delta\mathbf{g}_t)\,\mathrm{d}\Gamma = 0. \quad (6)$$

In case of sliding the tangential stress $\mathbf{t}_{T_i}$ at scale i is derived as, see e.g. [42],

$$\mathbf{t}_{T_i} = \langle \mu \rangle_{i-1}(p_{N_i}, v_i^{\text{loc}})\,\epsilon_N |g_N| \frac{\dot{\mathbf{g}}_{T_i}^{sl}}{\|\dot{\mathbf{g}}_{T_i}^{sl}\|}. \qquad (7)$$

The stick state is not influenced by the friction coefficient and therefore calculated as $\mathbf{t}_{T_i} = \epsilon_T\,\mathbf{g}_{T_i}$ by using a penalty regularization.

By knowing the local tangential stresses the procedure can be executed on the next larger scale. The derivation of the analytical friction function is done for each scale by fitting the average stresses to the assumed local friction law, see (4). In each step the effects of all lower scales are included by the averaged friction coefficient $\langle \mu \rangle_i$. The friction law from the previous scale and the material damping at the active scale then add up to the current

---

[2] A typical example of the frictional contact law is presented in (41).

friction law Therefore the friction values increase on each scale until the largest wavelength is reached.

The local contact pressures depend on the current scale, they increase for the blocks at smaller scale since there the area is also smaller. Only a fraction of the rubber block bottom is in contact with the rough surface on each scale. Therefore the absolute value of the true contact area decreases at smaller wavelengths. This leads consequently to larger local pressures, because the same vertical forces have to be transmitted via smaller contact zones.[3]

3 Constitutive Model for Elastomers

Hysteretic friction is related to energy dissipation inside the material during repeated loading. For elastomers the inelastic part of the constitutive model leading to dissipation is based on a viscoelastic ansatz. Damage is neglected as we are not interested in wear, but in long-time periodic excitations and related effects. Additionally the velocity dependency of friction should be investigated, which makes a time dependent ansatz necessary.

The detailed derivation of the viscoelastic model and its finite element implementation are not part of this work. For the explicit depiction of related models, see [15, 19, 20, 34]. A model for incompressible materials with finite linear viscoelasticity was proposed by Haupt and Lion [11]. The following terms state the basic relations needed to formulate the constitutive behavior of an elastomer for finite linear viscoelasticity setting.

The deformation gradient is split multiplicatively into an elastic and an inelastic viscous part for each Maxwell element k

$$\mathbf{F} = \mathbf{F}_e^k \cdot \mathbf{F}_i^k. \tag{8}$$

The right Cauchy–Green-Tensors for the inelastic part is defined by

$$\mathbf{C}_i^k = (\mathbf{F}_i^k)^T \cdot \mathbf{F}_i^k, \tag{9}$$

$$\mathbf{C} = (\mathbf{F}_i^k)^T \cdot \mathbf{C}_e^k \cdot \mathbf{F}_i^k, \tag{10}$$

and the evolution equations for the inner variables $\mathbf{C}_i^k$ are calculated with

$$\dot{\mathbf{C}}_i^k = \frac{1}{\tau^k}(\mathbf{C} - \mathbf{C}_i^k). \tag{11}$$

The strain energy function consists of a sum of the elastic energy and the strain energy related to the Maxwell elements, which is assumed to have the same form as the elastic strain energy

---

[3] Reasonable loading conditions are defined by a preliminary estimation of the true contact areas from the largest to the smallest scale, see Section 6.4.

$$\Psi_\infty = \frac{\mu_\infty}{2}(I_\mathbf{C} - 3 - 2\ln J) + \frac{\Lambda_\infty}{4}(J^2 - 1 - 2\ln J), \tag{12}$$

$$\Psi_{MW}^k = \frac{\mu^k}{2}(I_{\mathbf{C}_e^k} - 3 - 2\ln J_e^k) + \frac{\Lambda^k}{4}((J_e^k)^2 - 1 - 2\ln J_e^k), \tag{13}$$

$$\Psi = \Psi_\infty + \sum_k \Psi_{MW}^k. \tag{14}$$

The differentiation of the strain energy with respect to the Cauchy–Green strain tensor yields the second Piola–Kirchhoff

$$\mathbf{S} = \mathbf{S}_\infty + \sum_k \mathbf{S}_{MW}^k \tag{15}$$

with

$$\mathbf{S}_\infty = 2\frac{\partial \Psi_\infty}{\partial \mathbf{C}} = \mu_\infty(\mathbf{1} - \mathbf{C}^{-1}) + \frac{\Lambda_\infty}{2}(J^2 - 1)\mathbf{C}^{-1}, \tag{16}$$

$$\mathbf{S}_{MW}^k = 2\frac{\partial \Psi_{MW}^k}{\partial \mathbf{C}} = 2\frac{\partial \Psi_{MW}^k}{\partial \mathbf{C}_e^k} : \frac{\partial \mathbf{C}_e^k}{\partial \mathbf{C}} = 2(\mathbf{F}_i^k)^{-1} \cdot \frac{\partial \Psi_{MW}^k}{\partial \mathbf{C}_e^k} \cdot (\mathbf{F}_i^k)^{-T}$$

$$= \mu^k((\mathbf{C}_i^k)^{-1} - \mathbf{C}^{-1}) + \frac{\Lambda^k}{2}((J_e^k)^2 - 1)\mathbf{C}^{-1}. \tag{17}$$

A finite element implementation of the viscoelastic constitutive equation requires the algorithmic consistent tangent matrix which derivation is rather extensive. A detailed derivation of this matrix for finite linear viscoelasticity is presented in [33]. Its extension to six parallel Maxwell elements leads simply to the summation of the single Maxwell contributions.

In this contribution the considered material is an unfilled Styrene Butadiene Rubber (SBR). After measuring stresses and strains for varying loading frequencies, the rheometer identifies a frequency-dependent storage and loss modulus. Figure 2 depicts the frequency dependency in a double-logarithmic drawing.

Fitting the moduli with six Maxwell elements leads to a good approximation of the measured values, as depicted in the right-hand part of Figure 2. Note that the approximated loss modulus possesses a second maximum at an angular frequency $\omega \approx 10^9$ 1/s, which does not match the experimental curve. Indeed, the interesting frequency range in this work is limited by $\omega \approx 6 \cdot 10^6$ 1/s for a maximum velocity $v = 20.000$ mm/s at a wavelength $\lambda = 21$ μm. This indicates that the chosen model is sufficient, and a reduction of the error by using more Maxwell elements is not necessary.

The model parameters which are calculated from the measured data are listed in Table 1.

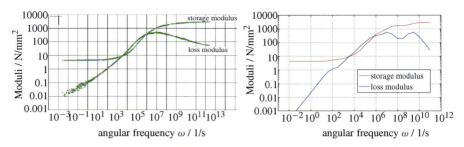

Fig. 2 Storage and loss modulus of chosen material SBR, measured values and approximation.

Table 1 Fitted material parameter SBR

MW element	$[E] = \text{N/mm}^2$	$[\tau] = 1/\text{s}$
–	4.17	–
MW1	1.72	0.01134034
MW2	7.36	0.0002628
MW3	70.98	1.316e-05
MW4	505.87	1.295e-06
MW5	1125.85	7.708e-08
MW6	1185.94	4.2e-10

4 Rough Surface Description

Many fractal surfaces like road tracks show self-affine behavior and thus have similar roughness characteristics on different length scales. A typical example of a two-dimensional self-affine track is a simple random walk (Brownian motion). It can be created by a random unit up and down movement of the function value z at each unit step on the x-axis. The roughness of this track is described by the Hurst-Exponent $H = 0.5$. The Hurst-Exponent or the related fractal dimension D ($D = \delta - H$ for the number of dimensions δ) denotes a measure for the surface irregularity, see [21]. $D = 2$ describes a smooth surface while $D = 3$ indicates the limit of an infinite rough surface in three dimensions.

In addition to this roughness parameter, two further values are necessary to describe the upper cut-off-point beyond which the self-affine properties start. These are the correlation lengths, parallel ($\xi_{\parallel}$) and normal ($\xi_{\perp}$) to the surface. If the horizontal distance of two points is larger than $\xi_{\parallel}$, their heights are independent. The correlation length $\xi_{\perp}$ is related to the statistical variance and therefore a measure for the root mean square fluctuations around the mean height

$$\xi_{\perp}^2 = 2\langle (z(x) - \langle z \rangle)^2 \rangle, \tag{18}$$

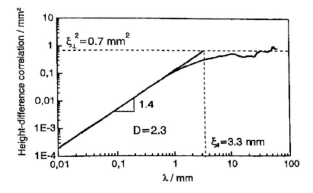

Fig. 3 Example of a height correlation function, see [17].

in which $\langle \ldots \rangle$ denotes the average value defined in (2). A surface can be described by the height-correlation function

$$C_z(\lambda) = \langle (z(x+\lambda) - z(x))^2 \rangle. \tag{19}$$

This function compares the height z of two nearby points with a horizontal distance λ. This height difference is averaged over all x-values. In case of small λ the mean height difference correlates to the horizontal gap, but beyond the distance $\lambda = \xi_{||}$, C_z levels out to a constant value. For self-affine surfaces $C_z(\lambda)$ can be written as

$$C_z(\lambda) = \left(\frac{\lambda}{\xi_{||}}\right)^{2H} \xi_\perp^2 \quad \text{for} \quad \lambda < \xi_{||}, \tag{20}$$

$$C_z(\lambda) = \xi_\perp^2 \quad \text{for} \quad \lambda > \xi_{||}. \tag{21}$$

An example of this function is depicted in Figure 3. The progress of the curve is linear on a double-logarithmic scale for small λ, in the further run C_z stays approximately constant. The exponent $2H$ denotes the slope of the gradient section.

For a more detailed description of self-affinity and common methods to express fractals, see [17] and the references therein.

4.1 Sine Wave

The chosen approximation function for the fractal surface are harmonic sine waves which are superposed to match the rough surface track characteristics. One sine function is defined by its amplitude a and wavelength λ, respectively the argument b

$$z(x) = a\sin(bx) \quad \lambda = \frac{2\pi}{b}. \tag{22}$$

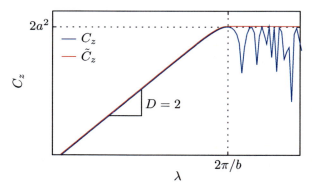

Fig. 4 Height correlation function of a simple sine wave.

The corresponding height correlation function is built using (19) to

$$C_z(\lambda) = \frac{b}{2\pi} \int_0^{2\pi/b} (a\sin(bx + b\lambda) - a\sin(bx))^2 \mathrm{d}x = 2a^2 \sin^2\left(\frac{b\lambda}{2}\right). \qquad (23)$$

Figure 4 shows that the general properties of a track approximated by sinusoidal functions are similar to the fractal surface. On a double logarithmic scale both functions depict a constant tangent for small λ until a maximum value around the upper cut-off length is reached. Hence the height values of two points at a larger interval are independent from each other. The oscillations, in case of the sine wave, beyond the cut-off point are due to the regularity of the function. Two points with a horizontal distance, which is a multiple of the wavelength, have always a height difference of zero.

However this relation has no influence on the height correlation function. The height difference variation of two points in a distance larger than the upper cut-off length of one wavelength, $\lambda = \xi_{||} = 2\pi/b$, is therefore assumed to be constant. Based on these observations the idealization of the height correlation function of the sine wave follows to

$$\tilde{C}_z = 2a^2 \sin^2\left(\frac{b\lambda}{2}\right), \quad \lambda \le \frac{2\pi}{b}, \qquad (24)$$

$$\tilde{C}_z = 2a^2, \quad \lambda > \frac{2\pi}{b}. \qquad (25)$$

For small λ Equation (23) yields

$$\bar{C}_z = \frac{a^2 b^2}{2}\bar{\lambda}^2, \qquad (26)$$

hence the gradient in the double logarithmic diagram is $2H = 2$. Changes of amplitude and wavelength affect the position of the first maximum point,

Fig. 5 Height profile of corundum 400 surface/μm.

but not the gradient of the height correlation function. After extension to three dimensions, the fractal dimension yields $D = 2$, which corresponds to a perfectly smooth surface and is smaller than typical values for track surfaces. Those are stated in [37] to be in the range of 2.1 to 2.4. A larger fractal dimension, respectively a smaller gradient of the slope, can be approximated by superposition of multiple sine waves with different amplitudes and wavelengths.

4.2 Application of the Approximation to a Rough Surface

The general similarity of the height correlation functions of a typical road track and a simple sine wave was shown in the previous section. A single sine function is however a bad approximation for a rough surface like the one depicted in Figure 5. To achieve a smaller gradient for the height correlation function several sine waves will be superposed. Two different sets of sine waves are introduced to approximate a corundum surface, the first one consists of just three harmonic functions, in the second set a smaller wavelength is added to the first one. This comparison is chosen in order to estimate the sensitivity of the used method with respect to the number of approximation functions. The amplitude to wavelength ratio of each sine wave is limited to ensure smoother surfaces, less local deformation. With these assumptions the numerical simulations can be performed in a robust way. Furthermore it was observed that larger amplitudes did not increase the accuracy of the fit, which depends mainly on the number of superposed approximation functions.

Figure 6 shows some superposed harmonic functions to illustrate what kind of roughness can be achieved with the chosen approximation. The second picture depicts a zoomed area of the first one and shows the similarity of different scales. In the third illustration (second zoom) the smallest sine wave of the approximation occurs and the function gets definitely smoother.

As real experiments were performed on a corundum surface (graining 400), this surface is used in the numerical tests. Corundum is one of the hardest minerals and therefore an established abrasive medium. Since wear effects are excluded for both contact partners, the material serves as a rigid unchanging regular surface with clearly measurable roughness properties. The largest wavelength of the corundum surface is in the range of 0.2 mm.

The profile was measured by the *Institute of Dynamics and Vibration Research, Leibniz Universität Hannover*. All track measurements were done

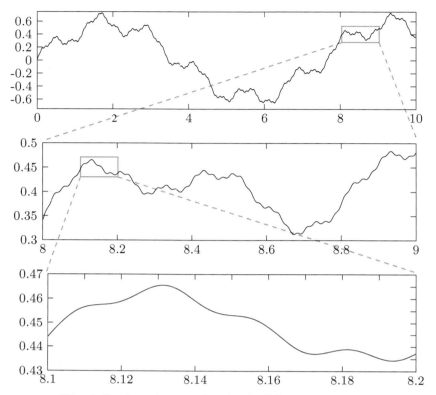

Fig. 6 Road track approximation in different zoom states.

along an arbitrary line of the surface with a minimal measuring distance of 7.5 μm.

4.2.1 Approach I – Three Sine Waves

The first approximation results from the superposition of three sine waves with different amplitudes and wavelengths. The height correlation functions of the sine waves and their superposition (red line) are depicted in Figure 7. The black crosses denote the measured values of the corundum surface for comparison. The gradient of each particular graph is obviously identical, but their cut-off-lengths differ due to applied amplitude and wavelength. Therefore the height-correlation of the added sine waves shows a smaller slope than the solitary functions and fits much better to the original. The fluctuations of the sine waves after their first maximum (compare Figure 4) are ignored to improve the clarity of the diagram.

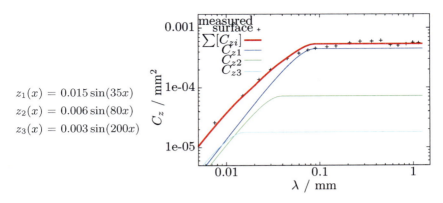

Fig. 7 Approximation of corundum 400 surface – surface I, 3 scales.

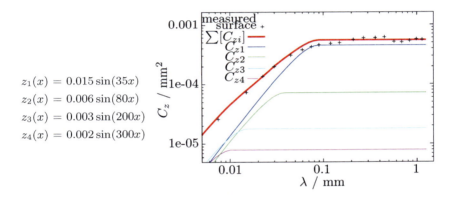

Fig. 8 Approximation of corundum 400 surface – surface II, 4 scales.

4.2.2 Approach II – Four Sine Waves

The second surface approach uses an additional "micro" sine wave. As the smallest wavelength in the first approach is about 31 µm, the influence of finer roughness has not been considered yet. Therefore an extra wavelength of $\lambda_{\min} \approx 21$ µm is added to the approximation functions of approach I. Figure 8 shows the single and the superposed sine approximations for this example. The black crossed line denotes the results of surface I. Only small percebtible variances are obtained in both height correlation functions. Contrary the numerical calculations on those surfaces show significant differences in the resulting friction values for both approaches which is due to the high influence of the micro scale roughness.

As a spatial FE model is used, the approximation functions are extended to three dimensions. Since the real roughness of surfaces does not have any distinct direction, an isotropic approximation is valid and the surface functions can be formulated in a product form

$$z(x,y) = a\sin(bx) \cdot \sin(by), \qquad (27)$$

using the coefficients a and b calculated for the two-dimensional functions. This formulation is sufficient, if the measured two-dimensional path is long enough to cover the maximum asperities of the profile and thus is representative for the surface. Since the cut-off point is passed and a constant height correlation is reached equation (27) fulfils these requirements.[4]

5 Contact

The geometrical constraints and the constitutive equations related to the contact zone which are needed in the weak form (6) are formulated in detail in this section. Beside standard normal and tangential contact, adhesional effects and heat conduction can play a potentially decisive role. This section deals with the basic relations concerning normal and tangential contact of a deformable body on a rigid analytical surface. The main derivation of kinematics and interface constitutive equations are well known and can be found in the literature, see [14] for analytical solutions, and [18, 42] for the computational approach to contact.

The enhancement of the contact formulation with adhesion will be added separately in Section 7. Heat fluxes at the contact zone are not covered in this work, for the coupled formulation we refer to [44].

5.1 Contact Kinematics and Interface Constraints

Each contact simulation starts with the search and identification of active contact zones. Therefore the precise description of the potential contact partners and their kinematical relations are important. In case of the contact of a deformable body with a rigid surface the reference surface Γ_c is chosen to be the rigid surface and called master surface. The detection of contact is relatively simple in that case. Using the description of the rigid surface, see e.g. (27), the analytical representation of the rigid surface is known.

Normal Contact

As already discussed, the fixed rigid surface is selected as reference surface and called "master" surface. The points on the deformable elastomer are denoted "slave" points. Hence all geometric values are defined on the rigid master surface and therefore the tangential vectors $\mathbf{a}_x^m, \mathbf{a}_y^m$ and the normal

---

[4] Further measurements on a road track are expected to give approximately equivalent results. The asphalt characteristics are similar in both directions as long as abrasion effects of the road as a result of one-way loading are neglected.

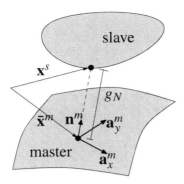

Fig. 9 Contact kinematics.

vector $\mathbf{n}^m$ can be calculated analytically at each point, see Figure 9. For simplicity tangential vectors are used which are located in the planes $x = 0$, respectively $y = 0$[5]

$$\mathbf{a}_x^m = \bar{\mathbf{x}}_{,x}^m = \begin{bmatrix} 1 \\ 0 \\ z_{,x} \end{bmatrix}, \quad \mathbf{a}_y^m = \bar{\mathbf{x}}_{,y}^m = \begin{bmatrix} 0 \\ 1 \\ z_{,y} \end{bmatrix} \tag{28}$$

$$\mathbf{n}^m = \frac{\mathbf{a}_x^m \times \mathbf{a}_y^m}{||\mathbf{a}_x^m \times \mathbf{a}_y^m||} = \frac{1}{\sqrt{z_{,x}^2 + z_{,y}^2 + 1}} \begin{bmatrix} -z_{,x} \\ -z_{,y} \\ 1 \end{bmatrix}. \tag{29}$$

The minimum distance between two points of the surfaces is defined as normal gap. The gap definition is the result of a closest point projection of a given slave point $\mathbf{x}^s$ onto the master surface. Generally the projection point $\bar{\mathbf{x}}^m$ is found by an iterative scheme since the representation (27) of the master surface is nonlinear

$$g_N \mathbf{n}^m = \mathbf{x}^s - \bar{\mathbf{x}}^m, \quad g_N \begin{bmatrix} n_1 \\ n_2 \\ n_3 \end{bmatrix}^m = \begin{bmatrix} x \\ y \\ z \end{bmatrix}^s - \begin{bmatrix} x \\ y \\ z(x,y) \end{bmatrix}^m. \tag{30}$$

This equation contains the three independent unknowns g_N, x^m and y^m. Inserting the third condition $g_N n_3^m = z^s - z^m$ into the other two yields the following expressions for the surface directions in the 3D case

$$G_x = x^s - x^m + z_{,x}^m z^s - z_{,x}^m z^m = 0, \tag{31}$$

$$G_y = y^s - y^m + z_{,y}^m z^s - z_{,y}^m z^m = 0. \tag{32}$$

---

[5] The specification of the tangents in x- and y-direction substitutes the usage of the convective coordinates $\xi\alpha$ to describe the contacting surfaces which are often found in standard contact formulations, see e.g. in [42].

A suitable procedure to solve this nonlinear system of equations is the Newton–Raphson-method.

The normal gap g_N can now be determined for each slave point. A penetration occurs if g_N is smaller than zero. Slave nodes which penetrate the master surface then build the active set for contact which has to be considered in the weak form (6). A gap larger than or equal to zero implies no active contact and provides no contact contribution to the system of equations. The contact regions and therefore the active set changes during the sliding process and must be updated in every computational step. Furthermore stresses in the contact area are less than zero in case of contact, or equal to zero in case of separation, $p_N \leq 0$. Both requirements lead to the Kuhn–Tucker–Karush conditions

$$g_N \geq 0, \quad p_N \leq 0, \quad g_N p_N = 0. \tag{33}$$

Hence either the stress or the normal gap is zero. This is only valid if adhesional effects are completely neglected. The extension to include adhesion in the normal and tangential contact theory is given in Section 7.

Tangential Contact

The tangential gap describes the relative movement of slave nodes parallel to the master surface. If two bodies stick together, no relative movement in tangential direction occurs between them, i.e. the tangential gap is zero. In case of sliding, the magnitude of the tangential gap and the path are not known a priori. These depend on the loading conditions and the constitutive friction equation. As the master surface is fixed in this case, the tangential gap is identical to the tangential change of the projection point $\bar{\mathbf{x}}^m$. The incremental path is defined along the tangent vectors of the master surface, see (28)

$$\mathrm{d}\mathbf{g}_t = \mathbf{a}_x^m \mathrm{d}x^m + \mathbf{a}_y^m \mathrm{d}y^m. \tag{34}$$

The total sliding distance of a slave point between time t_0 and t can be computed using $\mathrm{d}g_t = ||\mathrm{d}\mathbf{g}_t||$ and $\mathrm{d}x = \dot{x}\mathrm{d}t$, $\mathrm{d}y = \dot{y}\mathrm{d}t$

$$g_T = \int_{t_0}^{t} ||\mathrm{d}\mathbf{g}_t|| \, \mathrm{d}t \tag{35}$$

$$= \int_{t_0}^{t} \sqrt{\dot{x}^2(1+zx^2) + \dot{y}^2(1+zy^2)} \, \mathrm{d}t. \tag{36}$$

The tangential stress which can be transmitted between two contacting bodies is described by a constitutive model for friction. The classical law of Coulomb is a commonly used friction rule, which distinguishes between stick and slip state. Up to a certain tangential stress t_T, no tangential movement occurs and the bodies stick to each other. There is neither relative velocity nor movement between them, which is consequently written as

$$\dot{\mathbf{g}}_T^{st} = \mathbf{0} \Leftrightarrow \mathbf{g}_t^{st} = \mathbf{0}. \tag{37}$$

If the applied tangential load increases, the bodies start sliding. This leads to the formulation of a slip function

$$f_s = ||\mathbf{t}_T|| - \mu \, |p_N| \leq 0, \tag{38}$$

in which μ defines the friction coefficient. For $f_s < 0$ there is no relative tangential sliding and both bodies stick locally to each other. If the slip condition f_s is equal to zero, the contact state switches from stick to slip.

Friction forces act always in opposition to the moving direction and are colinear with the relative velocity vector. This follows from the slip rule, the slip function and the kinematical relations of the tangential gap, see e.g. [42]. This leads to

$$\mathbf{t}_T = -\mu \, |p_N| \frac{\dot{\mathbf{g}}_T^{sl}}{||\dot{\mathbf{g}}_T^{sl}||}. \tag{39}$$

In many practical applications the friction coefficient μ turns out to be constant and independent on contact area, temperature and sliding velocity. This assumption holds in most cases of rigid body contact but is not appropriate for rubber friction. Rubber sustains strong local deformations during sliding on a rough surface, and therefore the friction law turns out to be more complex. The most important influence on the frictional resistance is related to the sliding velocity and the normal stress. Thus (39) has to be modified to

$$\mathbf{t}_T = -\mu(v, p_N) \, |p_N| \frac{\dot{\mathbf{g}}_T^{sl}}{||\dot{\mathbf{g}}_T^{sl}||}. \tag{40}$$

The explicit form for the friction function depends on the material and surface properties. An example is given with Equation (41).

6 Numerical Results

Hysteretic friction is generally supposed to be the main factor of rubber friction. Following the algorithm in Section 2, the numerical calculations will start on the microscale without any local friction. The resulting friction function, depending on sliding velocity and normal pressure is transmitted to the next larger scale, where the procedure recurs. As already mentioned, the frictional resistance stems completely from the internal damping of the viscoelastic material.

The results depend on the one hand on the material properties of the used elastomer, on the other hand the characteristics of the surface are an important factor. This is shown by comparison of the two approximations "surface I" and "surface II", described in Section 4.2.

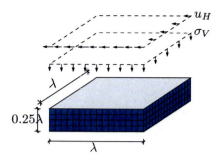

Fig. 10 System and loading.

6.1 System and Loading

A simple rectangle rubber block slides over a sinusoidal surface, which represents one scale of the harmonic approximation of a rough track, see Section 4. Due to the self-affine characteristics of the surface, it is possible to choose a similar finite element model for the rubber samples at different length scales.[6] Hence the problem definitions and the finite element model on each scale are similar and displayed in Figure 10. Only the sizes of sample and surface will change. The edge length of the block matches the wavelength λ of the respective current scale. The boundary conditions of the side nodes of the block are suitably linked. In this way, due to recursivity, the sample can be regarded as an infinite specimen parallel to the surface. Boundary effects like edge curling are excluded in this model. The finite element model for the elastomer solid and the details of the finite element contact formulation can be found in [35].

First, the block is loaded at its upper side by a prescribed vertical stress σ_V, which stays constant during the sliding process. Afterwards the upper side of the block is moved horizontally with a constant velocity $v := u_H/\mathrm{d}t$. The sliding distance u_H is always two sine waves long; $\mathrm{d}t$ is the size of the timestep.

Figure 11 depicts the vertical stresses in the contact zone together with the deformed mesh during a sliding process. During the first load steps the block deforms through shear and most contact elements still stick to their original position. Then the tangential contact forces increase and the sample starts sliding. After overcoming a slight maximum, the horizontal stresses settles at a nearly constant value. Since stationary conditions are needed to evaluate the stresses for homogenization the resulting stresses are only averaged over the second wavelength and then used to calculate the friction coefficient. According to Section 2, in the following we distinguish between the average stress $\langle p_N \rangle$ which is equal to the applied stress and the local pressure in the contact zone p_N. Note that the local pressure matches the global pressure on

---

[6] The general weak forms and finite element discretizations for viscoelastic rubberlike materials can be found in e.g. [43].

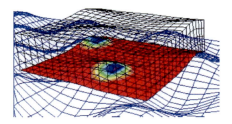

Fig. 11 Stresses in contact zone.

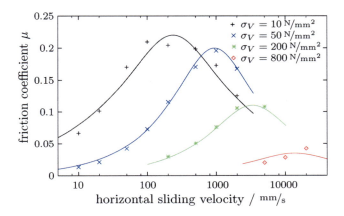

Fig. 12 Friction coefficient on microscale.

the next lower scale. Due to the decrease of the true contact area on smaller scales the mean local pressure increases immensely. Hence the applied global pressures used in the numerical tests go up to 800 N/mm^2.

6.2 Results on Microscale

The microscale results have been derived on the smallest scale of "surface II", i.e. the wavelength is about 0.021 mm with an amplitude of 0.002 mm.

The numerical experiments (see Figure 12) already show some important characteristics of the frictional behavior. Each dot in the diagram depicts the result of a finite element calculation, the solid lines are related to the approximation functions. It can be seen that the friction coefficient depends highly on the sliding velocity. For all observed normal stresses, the curves with respect to a logarithmic scale of the velocity on the x-axis are similar. This curves can be approximated by the function

Table 2 Fit parameters

σ_V N/mm^2	$\bar{v}$ mm/s	c —	$\mu^{\max}$ —	σ_H N/mm^2
10	236.7	0.41	0.22	2.2
50	943.1	0.65	0.20	10.0
200	3348.2	0.65	0.11	22.0
(800)	(13520)	(0.6)	(0.03)	(27.6)

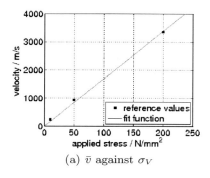

(a) $\bar{v}$ against σ_V

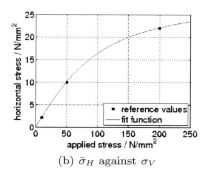

(b) $\bar{\sigma}_H$ against σ_V

Fig. 13 Position of the maximum points for different applied stresses on microscale.

$$\mu(v, \langle p_N \rangle) = \left(\frac{2v\bar{v}}{v^2 + \bar{v}^2} \right)^c \mu^{\max}(\langle p_N \rangle), \qquad (41)$$

in which $\bar{v}$ and $\mu^{\max}$ denote the maximum point of each curve. The broadening is controlled by the exponent c, which is assumed to be constant on each scale, because slight modifications of this parameter do not have significant effects.

The coordinates of the maxima depend on the applied pressure

$$\mu^{\max} = \mu^{\max}(\langle p_N \rangle) \quad \text{and} \quad \bar{v} = \bar{v}(\langle p_N \rangle). \qquad (42)$$

A decrease of the maximum friction value and a shift to larger velocities is observed for rising pressure. The values $\mu^{\max}$ and $\bar{v}$ are tabulated for the microscale in Table 2. The largest applied stress of 800 N/mm^2 gives only very few results and thus is not taken into account for the fitting procedure, but nevertheless they confirm the calculated fit parameters (dotted line in Figure 12).

As the viscoelastic model consists of multiple Maxwell elements with different relaxation times, the value of $\bar{v}$ is not unique, instead the "active" damping mode depends on the normal stress. The movement of the maximum point to larger velocities for increasing pressure can be explained as a result of the specific material parameters. Small spring stiffnesses of the used Maxwell elements correspond for this material to large relaxation times and vice versa, see Table 1. In case of high normal stress, the main part of the loading is therefore carried by the elements with larger stiffness. Because of the smaller relaxation times of these elements, the maximum of the friction coefficient moves to the right on the frequency axis.

The reference velocity for the different normal stress states is therefore related to the material properties. However it is complicated to propose an approximation function related to the material values. The crucial value is not the applied constant normal stress, but the local pressure distribution in the contact zone which is unknown. With sufficient precision the reference velocities can be expressed by a linear function, see Figure 13(a),

$$\bar{v} = a \langle p_N \rangle. \tag{43}$$

In Figure 12 the maximum friction coefficient seems to be similar for the smaller pressures of 10 and 50 N/mm^2, though larger pressures yield a strong decrease of $\mu^{\max}$. More evident than the friction coefficient is the horizontal resisting force for each loading case. As one would expect, the maximum horizontal stress grows with increasing normal stress. This dependency is nonlinear and the horizontal stress converges to a limiting value in case of perfect contact which is related to a very high pressure in normal direction. Figure 13(b) shows the approximation function for σ_H

$$\sigma_H = b \arctan(d \langle p_N \rangle) \tag{44}$$

$$\Rightarrow \quad \mu^{\max} = \frac{b}{\langle p_N \rangle} \arctan(d \langle p_N \rangle). \tag{45}$$

Altogether the friction law on microscale requires a fit of the four parameters a, b, c and d. This inverse problem is solved by a nonlinear least-square algorithm. The obtained homogenized friction law is then applied within each contact element at the next larger scale, where the local pressure p_N becomes the averaged contact pressure $\langle p_N \rangle$.

Since the finite element solution of the related weak form (6) is based on Newton's method, the partial derivatives of μ have to be computed within the friction part. They can be deduced from Equation (41)

$$\frac{\partial \mu}{\partial v} = c \left(\frac{2v\bar{v}}{v^2 + \bar{v}^2} \right)^{c-1} \frac{2\bar{v}(\bar{v}^2 - v^2)}{(v^2 + \bar{v}^2)^2} \mu^{\max}, \tag{46}$$

$$\frac{\partial \mu}{\partial p_N} = c \left(\frac{2v\bar{v}}{v^2 + \bar{v}^2} \right)^{c-1} \frac{2v(v^2 - \bar{v}^2)}{(v^2 + \bar{v}^2)^2} a \, \mu^{\max}$$

$$+ \left(\frac{2v\bar{v}}{v^2 + \bar{v}^2} \right)^c \frac{1}{p_N} \left(\frac{b\,d}{1 + d^2 p_N^2} - \mu^{\max} \right). \tag{47}$$

Thus all terms to build the consistent linearization of the friction coefficient are known. However, the magnitude and shape of the described properties depend directly on the used rubber material. Other compounds may generate completely different effects at least on the microscale and thus may change the dependencies in Equation (41).

6.3 Meso- and Macroscopic Results

The results on the next scales which yield the transition form micro- via meso- to macro-scale friction laws are now computed using the friction laws deduced from the previous scales. Figures 14(a) and 14(b) show the intermediate results on the larger scales of "surface II". The applied pressure values

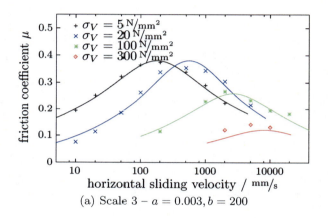

(a) Scale 3 – $a = 0.003, b = 200$

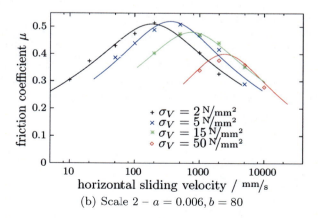

(b) Scale 2 – $a = 0.006, b = 80$

Fig. 14 Friction coefficient on larger scales.

get smaller on the larger scales due to the expected smaller local pressures on the larger wavelengths.

In general the friction values follow the same pattern as the microscale results. For each pressure the calculation series reaches a maximum friction value, whose magnitude decreases with increasing pressure. Furthermore the friction coefficient μ_i increases due to the additional local friction from the previous scale.

The sliding velocity at which the maximum friction occurs shifts its location for the same pressure on larger scales. This shift is related to the longer wavelength of the surface which requires a larger velocity to reach the same excitation frequency. The curves broaden on each scale, because the effects of the excitation frequencies interfere. The results of the numerical simulations on the largest scale of "surface II" ($\lambda \approx 0.18$ mm, $a = 0.015$ mm) are depicted in Figure 15 (solid line). They are closer to the experimental observations as the results of "surface I" (dashed line), an approximation neglecting the smallest scale, see Section 4.2. Even though the differences in the height correlation functions of the surface approximations are marginal, the effect on the friction results is prominent. Other exemplary calculations have shown that different choices of intermediate length scales lead to comparable results. However the lower cut-off-length is very influential. This result demonstrates clearly the importance of an exact surface description and of a correct detection of the lower cut-off-length.

The experimental results in Figure 15 were obtained at the *Institute of Dynamics and Vibrations (Leibniz Universität Hannover)* using a comparable rubber material on a corundum 400 surface. The numerical results for surface II are in good agreemnet with the measured values for small velocities, though the results diverge in case of higher relative sliding velocities. Unfortunately,

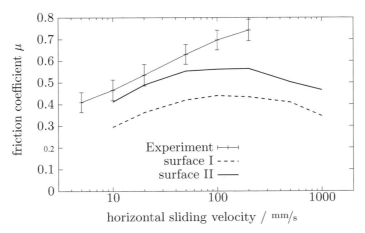

Fig. 15 Comparison experiment - simulation, $\sigma_V = 0.25$ N/mm^2.

experiments with higher speed could not be conducted for this material, hence the progression of the curve can only be guessed. Even the largest measured velocities may include side effects like wear due to damage in the comparatively weak rubber sample. On the other hand the numerical results vary slightly, due to chosen penalty parameters and mesh refinement. The latter becomes more important for increasing speed, because the smaller the true contact area is, the finer the contact zone has to be discretized.

Summing up the results show the applicability of the proposed multiscale method for hysteretic friction. The numerical calculations yield similar friction values for an exemplary rubber material on a rough surface and experimental laboratory tests. Hence the qualification of the chosen approach for the hysteretic energy dissipation on different scales is confirmed.[7] The computational results are very sensitive to the assumed conditions, more precisely the material parameters and surface characteristics. Hence, still extensive parameter studies have to be conducted to understand the quantitative connections. Howver the necessary numerical analysis for a sufficient large number of surfaces and rubber samples would be quite elaborate and would need parallel simulation techniques. This stems from the fact that each friction value results from a calculation with a great number of timesteps during the sliding process which leads to long computational times. Furthermore there is still a large amount of manual operations necessary which have to be performed during the multiscale transitions. An automation of these transitions would be desirable and thus has to be formulated in future work.

Additionally, a couple of important effects have not been incorporated so far. These are adhesional ad thermal effects. Formulations and examples for the influence of these phenomena are presented in Sections 7.3 and 8.2.

6.4 True Contact Area

The knowledge of the true area of contact of the footprint with the rough surface is for different investigations valuable. It can be obtained as a by-product of the previous investigations. The actual contact zone is supposed to be several magnitudes smaller than the macroscopic footprint, because the stress will in most cases not suffice to force the rubber inside the smaller cavities of the fractal surface. It influences the thermal fluxes and is often linked to adhesion as well. For a pure macroscopic model a good estimation of the true contact area leads to better models for heat transition or adhesion.

---

[7] The resulting friction curves show a distinct velocity dependence as well as a considerable connection to the applied vertical pressures. While the first result is consistent to experimental observations, the influence of pressure is so far widely neglected. This is probably due to limited pressure ranges of test rigs and arrangements such that large stress discrepancies were simply not tested.

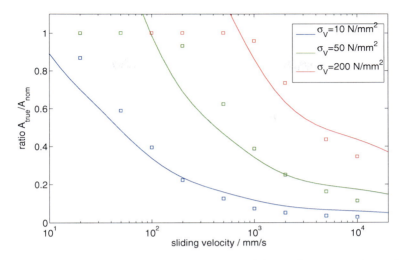

Fig. 16 Ratio of true contact area against sliding velocity on microscale.

Results on Smallest Length Scale

The true contact zone on the smallest length scale depends on the applied normal stress and the sliding velocity, which govern the stiffness of the rubber reaction. Due to the high local stresses at this scale, the ratio of the true contact area to the whole footprint of the sample is quite large. Even full contact is possible on this scale, compare the dotted results in Figure 16. As expected, the ratio increases with larger pressure and with smaller velocity which implies a weaker material behavior.

The numerical results are compared to the analytical Hertz solution for an elastic sphere on a rigid plane. The Hertz solution is based on the assumptions of a linear theory and thus can only approximate the real contact behavior. Within the theoretical framework the radius of the true contact area at an asperity can be written as

$$r = \sqrt[3]{\frac{3}{4}\frac{F(1-\nu^2)}{\kappa E^*(\omega)}}. \tag{48}$$

While the Poisson ratio ν is set to 0.5, the stiffness modulus E^* is assumed to change with the loading frequency ω. The curvature κ is chosen to match the reciprocal radius on the very top of the sine wave, which is described by the function $f(x,y) = a \sin bx \sin by$. This leads to $\kappa = ab^2$ and the normal load per contact point follows in this particular example from the applied vertical pressure to

$$F = \frac{1}{2} \sigma_V \lambda^2 \quad \text{with} \quad \lambda = \frac{2\pi}{b}. \tag{49}$$

The true contact area has the size $A_{true} = 2\pi r^2$. The nominal contact area corresponds to the sine wave surface area under the sample and is then

$$A_{\text{nom}} = \lambda \int_0^{2\pi/b} \left(\sqrt{1 + (ab \cos(bx))^2} \right) \, \mathrm{d}x. \tag{50}$$

The solid lines in Figure 16 depict the analytic ratio $A_{\text{true}}/A_{\text{nom}}$ for different stresses, while the dots denote the numerical results.

Considering the large discrepancies between analytical and numerical formulation, the results match surprisingly well. First, the curvature of the sine waves is not constant but decreases from top to mean axis, furthermore the filling of the cavities does not at all match the geometrical assumptions of Hertz. Second, the unsymmetric distribution of the true contact area in case of horizontal sliding (see Figure 17) differs extremely from the one of pure vertical pressure. Third, the Hertz solution is based on linear elastic material behavior. However all these errors seem to cancel out almost completely except for full contact, where the analytic Hertz solution becomes larger than one, which is unrealistic. Hence the analytical results can be used to obtain a macroscopic true contact area in the following calculations when sufficiently adjusted.

Macroscopic Results

The analytical microscopic results are now expanded to the larger scales. With a given pressure and the frequency-dependent stiffness the ratio of true contact area and footprint can be detected step by step. Each result

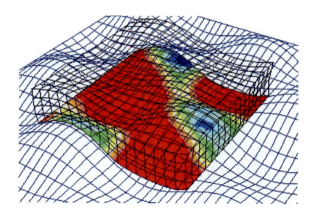

Fig. 17 Contact stresses in case of horizontal sliding.

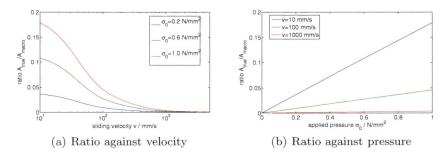

Fig. 18 Ratio of true contact area and macroscopic footprint, analytic approach.

leads directly to the area ratio on the next smaller scale, hence the mean pressure on this scale can be calculated. The true contact area follows from the multiplication of the area shares on each scale. It has to be kept in mind that the approximated stiffness is frequency-dependent and changes on different scales for the presribed velocity. Starting on the microscale with prescribed pressures and velocities, the contact areas A^1_{true} and A^1_{nom} can be calculated using Equations (48) to (50). The algorithm over the scales i is described with

$$p^i = \frac{A^i_{true}}{A^i_{nom}}, \quad \sigma^{i+1} = \frac{\sigma^i}{p^i}, \quad A^{i+1}_{true} = f(\sigma^{i+1}). \qquad (51)$$

Using four scales, the final true contact area is calculated as

$$A^4_{true} = p^1 \cdot p^2 \cdot p^3 \cdot p^4 \cdot A^1_{nom}. \qquad (52)$$

Figure 18(a) shows the ratio between true contact area and nominal footprint against the sliding velocity, which is in case of a macroscopic pressure of 6 bar smaller than 10%. The results match values in the literature, where the ratio is estimated to be about 1–2%, see [31]. The linear dependency of the true contact area on the applied pressure in Figure 18(b) has already been described by Archard [1] and Greenwood and Williamson [8].

Since the rubber reacts like a soft material for small sliding velocities, the local stresses are more evenly distributed and the real contact zone is larger than for increasing velocities. Extreme stress peaks occur for the largest loading frequencies, when just a few Gauss-points per asperity are in contact. Even in case of full contact, the contact zone is composed of areas with higher and smaller pressure. An example for the real pressure distribution is given in Figure 19.

These results are helpful for the deduction of macroscopic constitutive laws in the contact zone. If a detailed microscopic view is not necessary or too time-consuming, the estimation of the true contact area leads to a better approximation for adhesional stress or thermal flux between the surfaces.

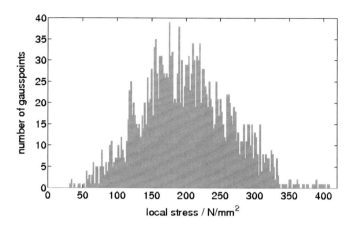

Fig. 19 Normally distributed pressures for $\sigma_V = 200$ N/mm^2, $v = 1000$ m/s on microscale.

7 Adhesion

The standard contact theory is based upon the Kuhn–Tucker–Karush conditions (33). While the non-penetration condition is physically motivated and still valid, the restriction to only compressive stresses in the contact interface is not completely true. Adhesion forces can be observed especially on very smooth surfaces. There exist several different models to include adhesion as well at micro-scale as at macro-scale level, see e.g. [7].

Here the model proposed by Raous, Cangémi and Cocu in 1999 [32] is applied. This approach includes adhesional forces both for the opening process of contact surfaces and the sliding motion with closed normal gap. Using the penalty approach, they introduce some additional parameters to describe the adhesional behavior. Basically the adhesion stiffness works analogously to the standard contact stiffness with inverse algebraic signs. Thus a penalty spring with the stiffness ϵ^a is introduced. The standard contact theory has to be changed by introducing an adhesion intensity β, see [7], which ranges between $\beta = 1$ for full adhesion and $\beta = 0$ for complete separation. During the seperation process the adhesion intensity will be maximal before the associated energy reaches the adhesive limit w. After that the adhesion parameter β starts to decrease. Due to the introduction of a viscosity parameter η the adhesive stresses may increase beyond this point, but will diminish to zero with growing gap and shrinking β. Two kinds of adhesion are defined: besides the adhesional tension forces, in case of opening of the gap there may as well occur an additional resistance force in horizontal direction during sliding. The theory asociated with the model [32] is presented next.

The free energy function Ψ^a for the adhesional case is written as

$$\Psi^a(\mathbf{g}^a, \beta) = \frac{1}{2}\epsilon^a||\mathbf{g}^a||^2\beta^2 - w\beta + I(\beta). \tag{53}$$

The adhesional gap $\mathbf{g}^a$ is defined as the vector between the current position of the slave point and the contact point $\overset{*}{\mathbf{x}}$ where the adhesion process started

$$\mathbf{g}^a = \mathbf{x} - \overset{*}{\mathbf{x}}. \tag{54}$$

The term $w\beta$ denotes the energy of decohesion and $I(\beta)$ is an indicator function which imposes the condition $\beta \in [0,1]$.

Partial derivatives with respect to the gap $\mathbf{g}^a$ and the adhesion intensity β yield the adhesional stresses σ^a and the thermodynamic force G_β

$$\sigma^a = \frac{\partial \Psi^a(\mathbf{g}^a, \beta)}{\partial \mathbf{g}^a} = \epsilon^a \mathbf{g}^a \beta^2, \tag{55}$$

$$-G_\beta = \frac{\partial \Psi^a(\mathbf{g}^a, \beta)}{\partial \beta} = \epsilon^a ||\mathbf{g}^a||^2 \beta - w \quad \text{if} \quad 0 \leq \beta \leq 1. \tag{56}$$

The rate of work of the adhesional stress has to be greater than the increase of adhesional energy, which implies a positive amount of energy dissipation during the process

$$\sigma^a \cdot \dot{\mathbf{g}}^a \geq \dot{\Psi}^a, \quad G_\beta \dot{\beta} \geq 0. \tag{57}$$

If G_β is positive or zero and

$$w - \epsilon^a ||\mathbf{g}^a||^2 \beta \geq 0, \tag{58}$$

the rate of adhesion intensity is automatically set to zero. The predefined factor β cannot increase. Hence the adhesion intensity will be held constant if (58) is fulfilled. Thus adhesion intensity β must consequently decrease such that $\dot{\beta} < 0$. The adhesion intensity is not reversible until contact is established again. Its evolution is chosen conforming to the thermodynamic force G_β under the influence of a viscosity parameter η which regards the observed time dependence of adhesional effects

$$\dot{\beta} = \frac{1}{\eta}\left(w - \epsilon^a||\mathbf{g}^a||^2\beta\right) \quad \text{if} \quad G_\beta < 0. \tag{59}$$

7.1 FEM

The finite element formulation of the chosen theory including derivation of the linearizations is presented in more detail in [25, 26]. At this point only the most important steps are presented starting with the discretization of the adhesional constitutive relation. The evolution equation (59) is integrated in time using the implicit Euler-Method which is consistent with the treatment of frictional contact

$$\dot{\beta}_{n+1} = \frac{\beta_{n+1} - \beta_n}{\Delta t} = \frac{1}{\eta}\left(w_{n+1} - \epsilon^a ||\mathbf{g}^a_{n+1}||^2 \beta_{n+1}\right). \tag{60}$$

Hence the adhesion parameter in the current timestep yields[8]

$$\beta_{n+1} = \frac{\eta\,\beta_n + \Delta t\,w_{n+1}}{\eta + \Delta t\,\epsilon^a ||\mathbf{g}^a_{n+1}||^2}. \tag{61}$$

The adhesional part of the weak form is similar to the standard contact theory. It results from variation of the energy potential (53) and integration over the adhesion area

$$G_a = \int_{\Gamma_a} \sigma^a \cdot \delta \mathbf{g}^a \, d\Gamma_a. \tag{62}$$

With the definition of the adhesional gap (54) the variation follows to $\delta \mathbf{g}^a = \delta \mathbf{u}$ and the discretized form of (62) assembled over n_a adhesion elements is obtained

$$G_a = \bigcup_{e=1}^{n_a} \sum_I \delta \mathbf{u}_I^T \int_{\Gamma_e} \mathbf{N}_I^T \sigma^a \, d\Gamma_e. \tag{63}$$

The integration area Γ_a is in case of tangential adhesion completely identical to Γ_c. If the gap opens the adhesion area is detected by several conditions: the normal gap has to be larger than zero, $g_N > 0$, but the observed point must have been in contact before. Additionally, the adhesion parameter β has to be still larger than a chosen tolerance close to zero, $\beta \geq TOL$. If β is beyond this limit adhesion will be neglected until standard contact has been established again.

During implementation one has to be careful handling the different contact states and the setting of history variables. Tension forces act only if the gap is open, while the additional sliding resistance is exclusively active in case of closed contact. If the gap status changes (new normal contact) the adhesion intensity β is set to one again. The displacement vector $\mathbf{g}^a$ denotes the adhesional gap depending on the respective contact point $\overset{*}{\mathbf{x}}$. In case of gap opening the last contact position is memorized and the adhesional gap is given by (54). More detailed descriptions can be found in [35].

The solution of the nonlinear finite element equations requires a linearization of the adhesional weak form (62)

---

[8] If η is set to zero, the time dependency cancels out and the new adhesion intensity can be calculated directly from (58) to $\tilde{\beta}_{n+1} = w_{n+1}/\epsilon^a ||\mathbf{g}^a_{n+1}||^2$ which leads to a pure function evaluation like in hyperelasticity.

$$\Delta G_a = \int_{\Gamma_a} \Delta \sigma^a \cdot \delta \mathbf{u} \, d\Gamma_a. \tag{64}$$

With

$$\Delta \beta_{n+1} = \frac{\partial \beta_{n+1}}{\partial \mathbf{u}} \cdot \Delta \mathbf{u} = \frac{-2\epsilon^a \beta_{n+1} \Delta t \, \mathbf{u}}{\eta + \Delta t \, \epsilon^a ||\mathbf{u}||^2} \cdot \Delta \mathbf{u} \tag{65}$$

the linearization of σ^a (55) yields

$$\Delta \sigma^a = \frac{\partial \sigma^a}{\partial \mathbf{u}} \cdot \Delta \mathbf{u} + \frac{\partial \sigma^a}{\partial \beta} \Delta \beta = \underbrace{\epsilon^a \beta_{n+1}^2 \left(1 - \frac{4\epsilon^a \Delta t(\mathbf{u} \otimes \mathbf{u})}{\eta + \epsilon^a \Delta t \, ||\mathbf{u}||^2}\right)}_{\Upsilon} \cdot \Delta \mathbf{u} \,. \tag{66}$$

These results lead to an additional tangent stiffness for adhesion

$$\Delta G_a = \bigcup_{e=1}^{n_a} \sum_I \sum_K \delta \mathbf{u}_I^T \mathbf{K}_{IK} \Delta \mathbf{u}_K \quad \text{with} \quad \mathbf{K}_{IK} = \int_{\Gamma_e} \mathbf{N}_I^T \Upsilon \mathbf{N}_K^T \, d\Gamma_e. \tag{67}$$

7.2 Adhesion Parameters

The model includes a number of material parameters, which are typically obtained by fitting procedures to experimental data. Unfortunately experiments which account for the velocity influence on adhesion are rare in the literature as adhesion measurements are often influenced by side effects. Consistent with the modelling so far the parameters are chosen based on microscopic considerations, more precisely the van der Waals interaction forces (e.g. [3, 16]).

The results are based on the stress between two flat surfaces

$$\sigma^a = \frac{A_H}{6\pi D^3}, \tag{68}$$

with the Hamaker constant A_H and the surface distance D. The Hamaker constant depends on the surface energy γ and the mean distance of two molecules D_0. Using a typical value for the surface energy of rubber $\gamma = 4.8 \cdot 10^{-5}$ Nmm, see [6], and a universal molecule distance of $D_0 \approx 0.165$ nm, see [3], the Hamaker constant follows as

$$A_H = 24\pi\gamma D_0^2 \approx 9.85 \cdot 10^{-17} \text{ Nmm.} \tag{69}$$

The adhesional work per unit area needed to separate the surfaces is given by

$$W^a = \int_{D_0}^{\infty} \sigma \, dD =: 2\gamma. \tag{70}$$

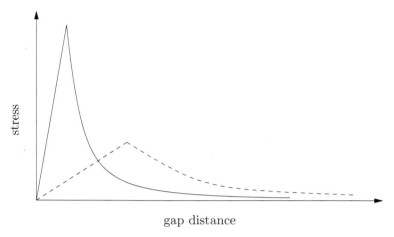

Fig. 20 Stress–gap relations for two different parameter sets.

Relation (68) denotes a correlation between adhesional stress and gap distance, neglecting any viscous influences. The parameters of the Raous adhesion model can be obtained by setting the viscosity η to zero. The work per unit area matches the limiting elastic energy w, which corresponds to $\epsilon^a \, ||\bar{\mathbf{g}}^a||^2$, see (58). $\bar{\mathbf{g}}^a$ denotes the distance where the adhesional stress (without viscosity) reaches its maximum. This distance should be in the magnitude of D_0, however this is far below the lowest length scale used in the calculations so far. Thus only the integral dissolving work is regarded here, which equals the limiting elastic energy $w = 2\gamma = 9.6 \cdot 10^{-5}$ Nmm. The increase of the open gap where adhesion is active leads consequently to a smaller maximum stress and a smaller adhesional stiffness to equal the performed work. Assuming a critical distance $||\bar{\mathbf{g}}^a|| = 0.5$ μm leads with

$$w = W^a = 2\gamma = \epsilon^a \, ||\bar{\mathbf{g}}^a||^2 \tag{71}$$

to an adhesional stiffness $\epsilon^a = 384$ N/mm^3. Figure 20 illustrates the approximation. Stress maximum and critical distance are varying, but the integral work, i.e. the area under the curves equals in both cases.

7.3 Numerical Results

This section extends the pure hysteretic approach of the calculations in Section 6 by including adhesion. For the adhesion model of Raous, Cangémi and Cocu the micromechanical motivated parameters, see last section, have been used. However this led to virtually no change of the pure hysteretic results

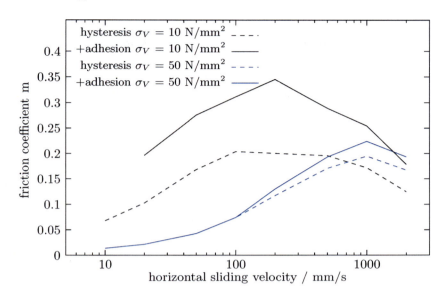

Fig. 21 Comparison of friction coefficient with and without adhesion.

reported in Section 6. The adhesion parameters are too small to produce a considerable effect. Hence, adhesion seems to have no significant influence on rubber friction, at least in this special application and the selected elastomer. However this conclusion is not supported by experimental observations, in which large differences of frictional behavior between wet and dry conditions are attributed to adhesional effects.

In the following the general operability of the model is presented by choosing larger adhesional parameters. The limiting elastic energy w of Section 7.2 is increased to $9.6 \cdot 10^{-3}$ Nmm which induces the same magnification for ϵ^a, if $||\bar{u}|| = 0.5$ μm still holds. Since adhesional effects are related to the real contact area, the computations were performed at the lowest scale ($a = 0.002$ mm, $\lambda = 0.021$ mm) under various loading conditions and for different sliding velocities. Figure 21 depicts the results for pure hysteretic friction as well as for friction including adhesional effects. The friction coefficient is plotted against the logarithmic sliding velocity for two vertical stresses, $\sigma_V = 10$ N/mm^2 and $\sigma_V = 50$ N/mm^2. A considerable increase of the friction coefficient is noticeable, especially for the smaller vertical stress. The reduced effect for higher vertical stresses is related to the increasing local normal contact stresses. In that case the tangential force is governed by the friction law due to hysteretic effects and the influence of the adhesional stress is less important for the friction coefficient μ.

8 Thermal Effects

In this section the influence of thermal effects on the hysteretic elastomer friction are investigated. Here the linear theory is used which however will be sufficient to show the thermal effects on elastomer friction. The unknown values in thermomechanical analysis are the total displacements $\mathbf{u}$ and the temperature Θ. On one hand the increase of Θ leads to thermal extensions and changing material properties, on the other hand the influence of energy dissipation during deformation on the system temperature is included.

8.1 Basic Equations

Temperature increase leads in general to a stress-free dilatation of the body. According to the linear theory, the strains are split in an additive manner into a mechanical ϵ^m and a thermal part ϵ^Θ. In case of a viscous or plastic material, the mechanical part is again divided into an elastic ϵ^e and an inelastic ϵ^i component

$$\epsilon^{\text{tot}} = \epsilon^m + \epsilon^\Theta = \epsilon^e + \epsilon^i + \epsilon^\Theta. \tag{72}$$

The thermal strain depends only on the temperature change and the expansion coefficient α_T. The kinematical relations lead then to

$$\epsilon^m = \epsilon^{\text{tot}} - \epsilon^\Theta = \frac{1}{2}(\text{grad}\,\mathbf{u} + \text{grad}^T\mathbf{u}) - \alpha_T(\Theta - \Theta_0). \tag{73}$$

Williams–Landel–Ferry-Transformation
In polymer science the validity of the principle of time-temperature superposition is generally accepted. It is known from experiments that a decrease of temperature and an increase of loading frequency have similar effects on the material properties of rubber. The material reacts stiffer in cold conditions and under larger loading frequencies. The relation between τ and Θ is used in experiments to widen the observable frequency scale, which is often strongly limited. Investigation of this small frequency band at different temperatures leads to an extension of the results. Figure 22 shows an example, where the measurements within a restricted frequency range at different temperatures are transferred to a broader frequency domain. Williams, Landel and Ferry [41] determined the shift factor a_t on the frequency axis in their WLF-transformation equation

$$\log a_z = \frac{-c_1(\Theta - \Theta_{\text{ref}})}{c_2 + \Theta - \Theta_{\text{ref}}}. \tag{74}$$

The equation is valid in a range from the glass transition temperature Θ_g to $\Theta_g + (70 - 100°\text{K})$. Williams, Landel and Ferry derived the universal material parameters referenced to the glass transition temperature Θ_g

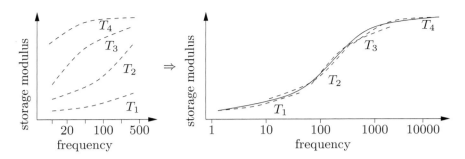

Fig. 22 Qualitative measuring results and master curve after WLF-shift.

$$c_1 = 17.44, \quad c_2 = 51.6. \tag{75}$$

For a standard tire elastomer a glass transition temperature is assumed to be $\Theta_g = -50°C$. Since the material stiffness and relaxation time τ_{ref} have been identified at room temperature a shift of the parameters to a reference temperature of $\Theta_{\text{ref}} = 20°C$ is necessary.

Equation (74) can be referenced to arbitrary temperatures by equating the shift factors in both directions [23]. The relation

$$\frac{-c_1(\Theta - \Theta_g)}{c_2 + \Theta - \Theta_g} = \frac{\bar{c}_1(\Theta_g - \Theta)}{\bar{c}_2 + \Theta_g - \Theta} \tag{76}$$

yields for $\Theta = \Theta_{\text{ref}} = 20°C$, including the condition $c_1 c_2 = \bar{c}_1 \bar{c}_2 = \text{const.}$, the aligned parameters

$$\bar{c}_1 = 7.4, \quad \bar{c}_2 = 121.6. \tag{77}$$

Thus the material change due to the temperature can be expressed for a viscoelastic Three-Parameter-model by a temperature dependency on the relaxation time

$$\tau = \tau_{\text{ref}} \cdot a_t(\Theta), \tag{78}$$

with τ_{ref} denoting the relaxation time at 20°C. Figure 23 shows a qualitative example for a shift of the friction curve due to application of a higher temperature. The WLF-transformation can reproduce the general thermomechanical relations of elastomers. If more realistic materials such as polymer blends are investigated, an additional vertical shift of the measurements could become necessary in certain ranges.

Generated Heat
During an inelastic deformation process a certain amount of energy is always dissipating out of the system. Observations show that this energy is not

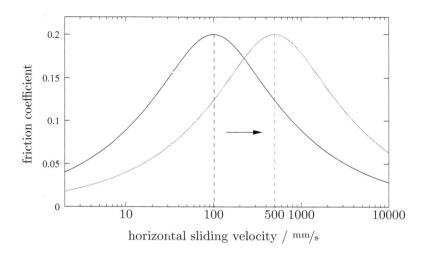

Fig. 23 Exemplary shift of the friction run in case of higher temperature.

lost, but acts like a heat source. Pure structural considerations ignore this heating, because either the heat quantity or its influences on the observed problem are small. However in frictional processes the generated heat can be quite substantial. The influence of the rising temperature on the elastomer properties was described in the last paragraph, the amount of heating depends on the viscoelastic deformation rate. We assume here a complete conversion of the dissipated viscous energy to heat. Therefore the heat quantity Q can be derived by

$$Q = \sigma : \dot{\epsilon}^i. \tag{79}$$

Finally the coupled weak forms for the mechanical and thermal problem are given by, see e.g. [44]

$$G^u = \int_{\mathcal{B}} \sigma : \mathrm{grad}^S \delta \mathbf{u} \; dV - \int_{\mathcal{B}} \rho \mathbf{b} \cdot \delta \mathbf{u} \; dV - \int_{\Gamma} \mathbf{t} \cdot \delta \mathbf{u} \; d\Gamma = 0, \tag{80}$$

$$G^\Theta = \int_{\mathcal{B}} \left(\rho c \dot{\Theta} \delta \Theta + k \, \mathrm{grad}^T \Theta \, \mathrm{grad} \delta \Theta - Q \delta \Theta \right) dV + \int_{\Gamma} \mathbf{q} \cdot \mathbf{n} \; \delta \Theta \; d\Gamma = 0. \tag{81}$$

8.2 Friction Test

The numerical calculations including thermomechanical coupling effects were performed using the following parameters:

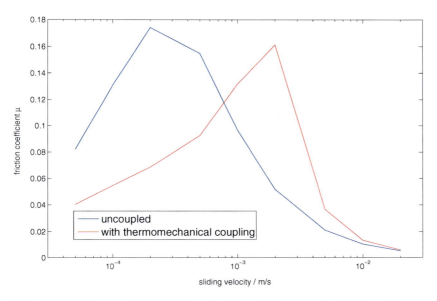

Fig. 24 Friction coefficient with and without thermomechanical coupling.

$$K = 4 \text{ MPa}$$
$$\mu^0 = 2 \text{ MPa}$$
$$\mu^\infty = 0.3 \text{ MPa}$$
$$\rho = 1200 \text{ kg/m}^3$$
$$\tau_{\text{ref}} = 1 \text{ s}$$
$$\Theta_0 = 20°\text{C}$$
$$\alpha_T = 0.0001 \text{ 1/K} \tag{82}$$
$$k = 0.3 \text{ W/mK}$$
$$c = 1600 \text{ J/kgK}$$

The numerical test concerns a viscoelastic block sliding over a sinusoidal surface, according to Section 6.1. Two kinds of linear viscoelastic elements are compared, on the one hand the fully coupled thermomechanical element and on the other hand an uncoupled version with the same material parameters, but unchanging relaxation time τ and no thermal expansion, i.e. $\alpha_T = 0$. The system is similar to Figure 10, using a wavelength and block width $\lambda = 0.1$ m and an amplitude of the sine wave of $a = 0.01$ m. The applied pressure has in this case a magnitude of $\sigma_V = 10^6$ N/m^2. The contact elements at the bottom of the block have no influence on the thermal behavior. Thus adiabatic heating conditions are assumed for the viscoelastic solid block. As a result, the rubber block will be continuously heated during sliding. The sliding distance is extended to fifty wavelengths, in contrast to the calculations in Section 6, where a sliding way of two wavelengths was sufficient.

Figure 24depicts the mean friction coefficient against different tested sliding velocities for the uncoupled and coupled material. The thermal influence causes a distinct shift of the friction curve on the logarithmic velocity axis. This matches the general predicted behavior in Figure 23 for rising temperatures. The changed form of the shifted curve is due to the different heating conditions for each velocity.

The chosen measurements of sine wave and block in this example are proportionally large, but tests with smaller wavelengths, amplitudes and accordingly smaller relaxation times τ lead to similar results. This example illustrates the large influence of the thermal conditions in and outside the rubber on the frictional behavior. It is assumed that an inclusion of the temperature effects will improve the results of Section 6.

9 Conclusions

The investigations have shown that it is possible to apply a multi-scale procedure to numerically compute the frictional behavior of elastomers on rough surfaces. In order to obtain good agreement with experimental observations the following ingredients have to be included in future work:

- Experiments which can be used to fit the constitutive parameters of the adhesion model in Section 7.
- Formulation of the thermo-mechanical coupled contact problem for large strains including heat transfer and eventually frictional heating at the contact interface.

References

1. Archard, J.F.: Elastic deformation and the laws of friction. Proceedings of the Royal Society of London, Series A 243, 190–205 (1957)
2. Bandeira, A.A., Wriggers, P., de Mattos Pimenta, P.: Homogenization methods leading to interface laws of contact mechanics. International Journal for Numerical Methods in Engineering 59, 173–195 (2004)
3. Bhushan, B. (ed.): Springer Handbook of Nanotechnology. Springer, Heidelberg (2004)
4. Bucher, F., Dmitriev, A.I., Ertz, M., Knothe, K., Popov, V.L., Psakhie, S.G., Shilko, E.V.: Multiscale simulation of dry friction in wheel/rail contact. Wear 261, 874–884 (2006)
5. Edwards, S.F., Vilgis, T.A.: The tube model theory of rubber elasticity. Reports on Progress in Physics 51, 243–297 (1988)
6. Diversified Enterprises, Solid surface energies, 101 Mulberry St. Suite 2N, Claremont NH 03743 U.S.A (2007),
http://www.accudynetest.com/surface_energy_materials.html
7. Frémond, M.: Adherénce des solides. Journal de Mécanique Théorique et Appliquée 6, 383–407 (1987)
8. Greenwood, J.A., Williamson J.B.P.: Contact of nominally flat surfaces. Proceedings of the Royal Society of London, Series A 495, 300–319 (1966)

9. Grosch, K.A.: The relation between the friction and viscoelastic properties of rubber. Proceedings of the Royal Society London, A 274, 21–39 (1963)
10. Haraldsson, A., Wriggers, P.: A strategy for numerical testing of frictional laws with application to contact between soil and concrete. Computer Methods in Applied Mechanics and Engineering 190, 963–978 (2001)
11. Haupt, P.: Continuum Mechanics and Theory of Materials. Springer, Heidelberg (2000)
12. Heinrich, G., Kaliske, M.: Theoretical and numerical formulation of a molecular based constitutive tube-model of rubber elasticity. Computational and Theoretical Polymer Science 7, 227–241 (1998)
13. Ihlemann, J.: Kontinuumsmechanische Nachbildung hochbelasteter technischer Gummiwerkstoffe, PhD Thesis, Universität Hannover (2003)
14. Johnson, K.L.: Contact Mechanics. Cambridge University Press, Cambridge (1985)
15. Keck, J.: Zur Beschreibung finiter Deformationen von Polymeren: Experimente, Modellbildung, Parameteridentifikation und Finite-Elemente-Formulierung, Dissertation, Universität Stuttgart (1998)
16. Kendall, K., Johnson, K.L., Roberts, A.D.: Surface energy and the contact of elastic solids. Proceedings of the Royal Society of London, Series A 324, 301–313 (1971)
17. Klüppel, M., Heinrich, G.: Rubber friction on self-affine road tracks. Rubber Chemistry and Technology 73, 578–606 (2000)
18. Laursen, T.A.: Computational Contact and Impact Mechanics. Springer, Berlin (2002)
19. Lion, A.: On the large deformation behaviour of reinforced rubber at different temperatures. Journal of the Mechanics and Physics of Solids 45, 1805–1834 (1997)
20. Loehnert, S.: An adaptive stepsize control for the integration of evolution equations in viscoelasticity, Diplomarbeit, Technische Universität Darmstadt (1999)
21. Mandelbrot, B.B.: The Fractal Geometry of Nature. W.H. Freeman, New York (1982)
22. Meier, J., Daemgen, M., Klüppel, M.: Micromechanics of rubber reinforcement by nano-structured fillers. Tagungsband des 6. Kautschuk-Herbst-Kolloquiums (2004)
23. Michaelis, H.: Thermomechanische Kopplung in der FEM. Universität Hannover, Bachelorarbeit (2005)
24. Mullins, L.: Effect of stretching on the properties of rubber. Journal of Rubber Research 16, 275–289 (1947)
25. Nitsche, R.: Numerische Untersuchungen zum Tragverhalten von gleitfest vorgespannten Schraubenverbindungen unter Berücksichtigung der Adhäsion. Universität Hannover, Diplomarbeit (2005)
26. Nitsche, R., Nettingsmeier, J., Rutkowski, T., Wriggers, P., Schaumann, P.: Numerische Berechnungen des Tragverhaltens von gleitfest-vorgespannten Schraubenverbindungen unter Berücksichtigung der Adhäsion. Der Bauingenieur 82, 77–84 (2007)
27. Ogden, R.W.: Large deformation isotropic elasticity: on the correlation of theory and experiment for incompressible rubberlike solids. Proceedings of the Royal Society of London, Series A 326, 565–584 (1972)

28. Orlik, J., Zhurov, A., Middleton, J.: On the secondary stability of coated cementless hip replacement: parameters that affect interface strength. Medical Engineering and Physics 10, 825–831 (2003)
29. Payne, A.R.: The dynamic properties of carbon black-loaded natural rubber vulcanizates, Part I. Journal of Applied Polymer Science 19, 57–63 (1962)
30. Payne, A.R.: The dynamic properties of carbon black-loaded natural rubber vulcanizates, Part II. Journal of Applied Polymer Science 21, 368–372 (1962)
31. Persson, B.: Theory of rubber friction and contact mechanics. Journal of Chemical Physics 115(8) (2001)
32. Raous, M., Cangémi, L., Cocu, M.: A consistent model coupling adhesion, friction and unilateral contact. Computer Methods in Applied Mechanics and Engineering 177, 383–399 (1999)
33. Reese, S.: Thermomechanische Modellierung gummiartiger Polymerstrukturen. Universität Hannover, Habilitationsschrift (2001)
34. Reese, S., Govindjee, S.: A theory of finite viscoelasticity and numerical aspects. International Journal of Solids and Structures 35, 3455–3482 (1998)
35. Reinelt, J.: Frictional contact of elastomer materials on rough rigid surfaces, PhD thesis, Leibniz Universität Hannover, Institute for Continuum Mechanics (2008)
36. Schallamach, A.: The velocity and temperature dependence of rubber friction. Proceedings of the Physical Society, Section B 66, 386–392 (1953)
37. Schramm, E.J.: Reibung von Elastomeren auf rauen Oberflächen und Beschreibung der Nassbremseigenschaften von PKW-Reifen, PhD thesis, Universität Regensburg (2002)
38. Simo, J.C.: On a fully three-dimensional finite-strain viscoelastic damage model: Formulation and computational aspects. Computer Methods in Applied Mechanics and Engineering 60, 153–173 (1987)
39. Stupkiewicz, S.: Micromechanics of Contact and Interface Layers. Springer, Berlin (2007)
40. Tworzydlo, W.W., Cecot, W., Oden, J.T., Yew, C.H.: Computational micro- and macroscopic models of contact and friction: Formulation, approach, applications. Wear 220, 113–140 (1998)
41. Williams, M.L., Landel, R.F., Ferry, J.D.: The temperature dependence of relaxation mechanisms in amorphous polymers and other glass-forming liquids. Journal of the American Chemical Society 77, 3701–3707 (1955)
42. Wriggers, P.: Computational Contact Mechanics. Springer, Berlin (2006)
43. Wriggers, P.: Nonlinear Finite Element Methods. Springer, Berlin (2008)
44. Wriggers, P., Miehe, C.: Contact constraints within coupled thermomechanical analysis – A finite element model. Computer Methods in Applied Mechanics and Engineering 113, 301–319 (1994)
45. Zohdi, T., Wriggers, P.: An Introduction to Computational Micromechanics. Springer, Heidelberg (2004)

Thermal Effects and Dissipation in a Model of Rubber Phenomenology

D. Besdo, N. Gvozdovskaya, and K.H. Oehmen

Abstract. In this article, the MORPH rubber model [1, 7] is expanded to thermal effects like changes in dissipation and elastic behaviour with rising temperatures, based on experimental data of the German Institute of Rubber Technology (DIK): uniaxial tension cycles at different amplitudes and temperatures with six different rubber compounds. The standard MORPH model depends on eight constant parameters which are changed to functions of temperature.

The second part deals with the distinction between reversible elastic power and irreversible dissipation. The dissipation rate must not be negative at any time and leads to heat production per volume which results in temperature rise. The rubber model MORPH deals only with stresses and strains, and only the dissipated work per cycle in periodic loading is fixed by the characteristic hysteresis curves, depending on amplitude and deformation history. Here, a formulation for an elastic energy density is presented, which results in non-negative dissipation power in most loading variants. The average dissipation of complete loading cycles is always positive.

1 Introduction

The rubber model MORPH (model of rubber phenomenology) [1, 7] is able to simulate characteristic hysteresis effects for small and large deformations, depending on the deformation history by a history invariant of the Cauchy–Green deformation tensors $\underline{C}$ or $\underline{b}$. So far, this rubber model depends on eight parameters describing the elastic and inelastic stress-strain relations due to previous and actual large deformations. Strong nonlinearities,

---

D. Besdo · N. Gvozdovskaya · K.H. Oehmen
Institute of Continuum Mechanics, Leibniz Universität Hannover, Appelstr. 11, 30167 Hannover, Germany
e-mail: {office, gvozdovs, oehmen}@ikm.uni-hannover.de

hysteresis and softening effects due to such deformations can be described by this model. The required parameters can be identified by comparing special experimental data of uniaxial tension or simple shear with calculations using this rubber model [3], but also tests with inhomogeneous strain distributions can be used to adapt the material constants [6]. It is developed to be used in finite element calculations working with small time steps. Many calculations have been performed by implementing this model in the finite element package ABAQUS and in our own finite element programs of the Institute of Continuum Mechanics, Leibniz Universität Hannover [1, 4, 5].

The stresses at a new time step can only depend on deformations and deformation velocities and their history values as well as on stresses at the last iterated time step. Frequencies and amplitudes of periodic cycles may not be input values for the constitutive model. Popular values like G_{eff}, G', G'' and $\tan \delta$ are no material constants for rubber, they depend on frequency, amplitude and mean deformation.

The stress-strain relations are time independent, so changing deformation velocity or changing from sinusoidal to constant loading and unloading velocity has no effect on the characteristic hysteresis curves.

After a short review of the standard MORPH model in Section 2, the thermal effects to dissipation and elastic behaviour are investigated in uniaxial tension experiments of the German Institute of Rubber Technology (DIK) in Section 3. From these results different phenomenological extensions of the MORPH model are derived defining the eight material parameters as functions of temperature. This ends up in doubling the constants, so up to 16 constants have to be adapted by least square sums of differences between experimental data and simulating calculations.

In periodic loading, the dissipated work per cycle is fixed by the characteristic hysteresis curves, depending on amplitude and deformation history. In experiments with uniaxial tension the hysteresis area in a diagram $T_{11}(F_{11})$ corresponds to the dissipated work per cycle and per volume. But nothing is said about the dissipation distribution during one load cycle. The second law of thermodynamics postulates a nonnegative dissipation rate at all times, and that shall be achieved by an energy density formulation including nonlinear elastic effects in the MORPH rubber model.

Thermal expansion, thermal conductivity, specific heat capacity and density of rubber are not considered within this investigation, for computations typical constant values are applied.

2 The Standard Rubber Model MORPH

This model [1, 7] applies a Tresca-like invariant of the deformation tensor $\underline{\underline{b}} = \underline{\underline{F}} \cdot \underline{\underline{F}}^T$ and the deformation rate tensor $\overset{*}{\underline{\underline{b}}}$, where $*$ denotes the Jaumann–Zaremba time derivative. b_T is the largest difference of the eigenvalues of the corresponding tensor $\underline{\underline{b}}$. The history invariant b_T^H is the maximum of b_T

in previous times. Both invariants are identical with those of $\underline{\underline{C}} = \underline{\underline{F}}^T \cdot \underline{\underline{F}}$, that is $C_T = b_T$ and $C_T^H = b_T^H$.

The model can be formulated in Euler or Lagrangian form [2, 7] and the output can be transformed to Cauchy stresses $\underline{\underline{\sigma}}$, to first Piola–Kirchhoff stresses $\underline{\underline{T}}$ or to second Piola–Kirchhoff stresses $\underline{\underline{\tilde{T}}}$, abbreviated as 1st or 2nd PK. The rubber compounds are considered to be nearly incompressible with a large bulk modulus $K > 2$ kN/mm², J represents the actual volume ratio and J_ϑ the volume ratio due to temperature change $\Delta\vartheta$:

$$\begin{aligned} J &= v/\tilde{v} = \tilde{\rho}/\rho \quad \text{actual volume ratio,} \\ J_\vartheta &= 1 + 3\alpha\Delta\vartheta \quad \text{with } \alpha = \text{thermal expansion coefficient.} \end{aligned} \quad (1)$$

The hydrostatic stresses are proportional to volume change, all further stresses are physically deviatoric – index D – that means

$$\text{Cauchy stress} \quad \underline{\underline{\sigma}}_D = \underline{\underline{\sigma}} - \frac{1}{3}(\underline{\underline{\sigma}} \cdot \cdot \underline{\underline{1}})\,\underline{\underline{1}} \qquad \text{with } \underline{\underline{\sigma}} = J\,\underline{\underline{F}}^{-1} \cdot \underline{\underline{T}},$$

$$\text{1st PK stress} \quad \underline{\underline{T}}_D = \underline{\underline{T}} - \frac{1}{3}(\underline{\underline{T}} \cdot \cdot \underline{\underline{F}})\,\underline{\underline{F}}^{-1} \quad \text{with } \underline{\underline{T}} = \frac{1}{J}\,\underline{\underline{F}} \cdot \underline{\underline{\sigma}}, \quad (2)$$

$$\text{2nd PK stress} \quad \underline{\underline{\tilde{T}}}_D = \underline{\underline{\tilde{T}}} - \frac{1}{3}(\underline{\underline{\tilde{T}}} \cdot \cdot \underline{\underline{C}})\,\underline{\underline{C}}^{-1} \quad \text{with } \underline{\underline{T}} = \underline{\underline{\tilde{T}}} \cdot \underline{\underline{F}}^T.$$

The stresses are split into three parts: hydrostatic stresses $\underline{\underline{\sigma}}^h$, deviatoric Neo-Hooke stresses $\underline{\underline{\sigma}}_D^{NH}$ and additional deviatoric stresses for nonlinear elastic and inelastic effects $\underline{\underline{\sigma}}_D^A$:

$$\begin{aligned} \underline{\underline{\sigma}} &= \underline{\underline{\sigma}}^h + \underline{\underline{\sigma}}_D^{NH} + \underline{\underline{\sigma}}_D^A && \text{with} \\ \underline{\underline{\sigma}}^h &= K\,(J - J_\vartheta)\,\underline{\underline{1}} && \text{and} \\ \underline{\underline{\sigma}}_D^{NH} &= 2\alpha_e\,\underline{\underline{b}}_D / J && \text{with} \quad \alpha_e = p_1 + p_2 \cdot f(p_3\,b_T^H) \quad \text{and} \\ \underline{\underline{\sigma}}^A &= \beta\,b_T\,(\underline{\underline{\sigma}}^L - \underline{\underline{\sigma}}^A) && \text{with} \quad \beta = p_4 \cdot f(p_3\,b_T^H) \quad \text{and} \quad (3) \\ \underline{\underline{\sigma}}^L &= \frac{\gamma}{J}\,\exp\left(p_7\,\frac{\overset{*}{\underline{\underline{b}}}\,b_T}{\overset{*}{b_T}\,b_T^H}\right) + \frac{p_8\,\overset{*}{\underline{\underline{b}}}}{J\,\overset{*}{b_T}} && \text{with} \\ f(x) &= 1/\sqrt{1+x^2} && \text{and} \quad \gamma = p_5\,b_T^H\,[1 - f(b_T^H/p_6)]. \end{aligned}$$

Here α is the linear thermal expansion coefficient and α_e the Neo-Hooke parameter, which depends on the deformation history in this MORPH rubber model. The additional stresses are defined by a differential equation of time, which results in an exponential approach of the additional stresses to a limiting stress $\underline{\underline{\sigma}}^L$. This limiting stress tensor is the sum of an exponential tensor function of the deformation rate tensor $\overset{*}{\underline{\underline{b}}}$ and of a normalised form of $\overset{*}{\underline{\underline{b}}}$

itself. The parameters p_3 and p_5 are responsible for the softening effects with growing values for the deformation history, concerning the elastic Neo-Hooke stresses, the factor for approaching the limiting stresses and the factor for the exponential part of the limiting stresses. The parameters p_1 and p_2 are used for the elastic Neo-Hooke stresses, p_4 is a factor for approaching the limiting stresses. Parameter p_8 stands for a constant part of the hysteresis height in diagrams $\sigma_{11}(F_{11})$ and parameters p_5 and p_7 form the exponential part of the limiting stresses.

This model with eight parameters will be called standard MORPH model, it has no temperature effects and deals only with stresses and strains.

3 Implementation of Thermal Effects

Now these eight MORPH parameters shall be functions of temperature with new constants for each parameter. The MORPH procedure for computing stresses shall be unchanged, but before entering this routine, the eight MORPH parameters are determined by these temperature functions.

In the following diagrams for uniaxial tension, the 1st PK stress T_1, representing tensional force per original cross section, is plotted versus the linear strain $\epsilon_{\text{lin}} = \lambda - 1 = L_M/L_0 - 1$, corresponding to the experimental data.

3.1 Experimental Data for Thermal Effects in Six Rubber Compounds

In 2005 and 2007, the German Institute of Rubber Technology (DIK) performed uniaxial tension experiments at four temperatures with six rubber compounds [8]. For each compound they vulcanised four identical specimen – dumbbells with a cylindrical center part of 15 mm diameter and 21 mm length – and carried out uniaxial tension experiments with one specimen at one temperature with many cycle groups of constant amplitudes, increasing from group to group. Each cycle group consists of five cycles with the same displacement limits $1 < \lambda < \lambda_a$. The older experiments were accomplished with rubber compounds named E6R and S65K at temperatures of 24, 50, 70 and 90°C, the newer experiments with rubber compounds named S60N1, S60N3, SB6R1 and N330 at temperatures of 3, 22, 50 and 80°C. The amplitudes varied from 2% to 100 or 130%.

The reference length L_0 is the distance between two reflection markers within the cylindrical part of the dumbbell test pieces before any loading is applied but already at the specified temperature, so the thermal expansion is already included in L_0. L_M is the distance between the same two reflection markers at the actual deformed configuration corresponding to the actual 1st PK stress.

In Figure 1 the results for all specimen at all measured temperatures of one rubber compound are shown in one diagram. The curves show the last

Thermal Effects and Dissipation in a Model of Rubber Phenomenology 99

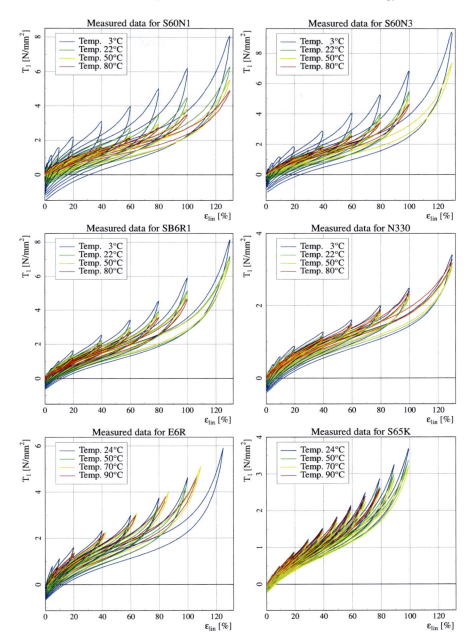

Fig. 1 Measured data for all six rubber compounds at four temperatures. 1st PK stresses T_1 are plotted versus linear strains ϵ_{lin} for a reduced number of base points of final cycles of each amplitude.

cycle of each amplitude in blue, green, yellow and red for rising temperatures. For rubber compound E6R the strains at high temperatures are recalculated by applying the listed values L_M and L_0 defined above. The listed strains are improperly related to the unstressed length after the last cycle group and not to L_0. For all compounds, the damping decreases with rising temperatures, the elastic behaviour is different for the six compounds. Mostly the elastic stiffness also decreases with rising temperatures, but for rubber compounds N330 and S65K the stiffness seems to stay nearly constant. The weakly damped behaviour at high temperatures leads to the effect that these hysteresis curves lie above the thick hysteresis curves at low temperatures, which show more negative stresses at positive strains.

3.2 Programs for Simulating Experiments and Identifying All Material Constants

The extension to thermal effects increases the number of material constants to be adapted and the amount of experimental data. That makes it necessary to develop a new package of separate programs:

- Calculating stresses by selected constitutive models at a given deformation is performed by special Fortran routines combined in a separate dynamic link library, the same as in our finite element systems for one integration point.
- A C-program named SpaDe simulates series of special deformation cycles calling the fortran library routines at any iteration at any time and organising graphical and listed output of stresses, strains, elastic energy and dissipation for further use. An xml-file controls the simulations by a sequence of commands for material type, deformation and deformation rate, and specifications for input and output.
- PID is a new program with two main tasks: the first to read the experimental data files of the DIK, to find automatically the last cycle of a group of cycles with the same amplitude. These cycles are essential for the parameter identification. Further on, this task prepares and saves all necessary steps before starting the optimisation thread. The second task is the parameter identification itself, which calls SpaDe with varying parameter sets. It can easily be restarted for new optimisations with another variant of thermal effects or with different starting values, relaxation or damping values for the optimisation process – without repeating the input and preparation task.

The optimisation of all material constants is performed by least square sum methods. There are several possibilities for weighting either each measure point or each cycle. The experimental data contain measuring points at constant deformation differences, for cycles with 2% strain about 20 points and for cycles with 100% linear strain about 1000 points. Constant weight per point would lead to very small influence of small cycles, but constant weight per cycle would lead to a large influence of a small hysteresis area. Here a reduction of measure points to n base points for least square sums is preferred, achieved by a special formula depending on the amplitude a in % of the cycle group:

$$n = \text{Max}[20, 10 \cdot \sqrt{a}] \quad \Longleftrightarrow \quad a \leq 4\% : n = 20, \quad a = 100\% : n = 100. \quad (4)$$

The first and last measure points of each cycle branch are always taken as base points, the stress and strain values of all other equally spaced base points are interpolated, the strain values can also be smoothed so that they are monotonically increasing in loading branches and decreasing in unloading branches.

An indirect weighting is also achieved by choosing the amplitudes of cycle groups. If the larger amplitudes vary in steps of 10 instead of 20%, the influence of the small cycles at 2 and 5% decreases by factor 2.

At the end of this section in Figure 9 different weights for different temperatures are considered, so that the errors are set relative to the average hysteresis height instead of a constant reference stress.

Another possibility would be to give different weight to the vertical position and to the height of the hysteresis curves. Then the stiffness or the damping gets higher influence at the expense of the other one, just as needed.

With regard to so many adaptations and to so many constants to be adapted, it is very important to have a large convergence radius and to control the changes of all parameters to get good convergence for the first iterations. In PID this can be achieved by keeping some parameters fixed for some iterations or by relaxation factors or by damping the adaptation procedure. It is also possible to set lower and/or upper limits for any parameter. Depending on the version of the constitutive model the number of parameters to be adapted can be chosen freely.

The parameter identification is based on the minimisation of the error square sum

$$\text{ESS} = \frac{1}{n} \sum_{i=1}^{n} (M_i - R_i)^2 \quad (5)$$

with M_i as interpolated measured value at the base point i and R_i as calculated value. For comparing the adaptations of all six compounds and all constitutive model variants, a mean error related to the stress at 100% strain ($\lambda = 2$) is defined in the form

$$mE[\%] = 100 \cdot \frac{\sqrt{\text{ESS}}}{M_{(\lambda=2)}}, \quad (6)$$

so the mean error is independent of hard and soft rubber compounds.

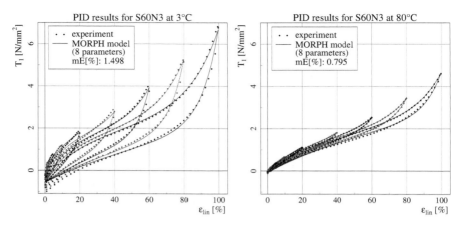

Fig. 2 MORPH adaptations for rubber compound S60N3 at 3 and 80°C.

3.3 Results for Thermal Effects in MORPH Model

With the new program suite, the experimental data of the DIK are analysed. At first each specimen at each temperature is considered as a special rubber compound, and the eight MORPH parameters are adapted to the data at one temperature. Without any problems the mean error can be minimised, starting with value 1 for all 8 parameters. The resulting mean error lies in the range of 0.5 to 2%, mostly around 1%. For the compound S60N3 the results are shown in Figure 2 for 3 and 80°C. The mean errors for 22 and 50°C are shown in Figure 7. For all compounds the largest mean error is found at the coldest temperature, due to deviations in the last section of the unloading branches of the hysteresis curves.

Now the changes for the 8 adapted parameters at all four temperatures are considered, in Figure 3 they are plotted in logarithmic scale. Regarding the trends of all 8 parameters, it seems to be possible to keep some of them constant in some compounds – but not in all six – and to use best fit straight lines with two constants in the logarithmic plots.

In a first attempt the constants are determined just by these four sets of six MORPH parameters without further optimisations. But this procedure is not satisfactory, since the sensitivity of the hysteresis curves to changes in these eight MORPH parameters is not taken into account and the set of eight parameters is not so completely independent as expected. If one parameter is changed by a small amount and then kept fixed, that leads certainly to a larger mean error if all other parameters are not changed, but when the other seven parameters are adapted to optimal values, the resulting mean error lies just slightly above the minimum value of all eight adapted parameters. Therefore the variation of the other seven parameters can nearly

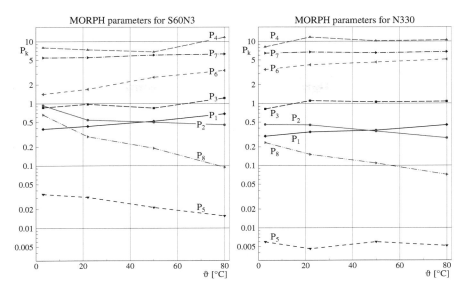

Fig. 3 MORPH parameters for rubber compounds S60N3 and N330 depending on temperature.

compensate a moderate change of one parameter, but the error distribution for all hysteresis curves has changed leading to nearly the same mean error.

Next the idea of best fit lines for the eight MORPH parameters is used to find thermal functions for all of them, requiring two new constants A_k and B_k for each MORPH parameter:

$$\text{variant 1:} \quad p_k = A_k \cdot \exp(B_k \, \Delta \vartheta) \quad \text{for} \quad k = 1 \ldots 8 \, . \tag{7}$$

Here, room temperature of 22°C is set as an arbitrary reference temperature. The values A_k represent the MORPH parameters at this reference temperature and the values B_k stay for the exponential dependency on temperature difference $\Delta \vartheta$ to this reference temperature. All A_k must be positive like p_k; $B_k = 0$ leads to a temperature independent parameter p_k, $B_k < 0$ decreases the MORPH parameter for rising temperatures, e.g. for $k = 2$, 5 or 8, whereas $B_k > 0$ increases the MORPH parameter, e.g. for $k = 1$, 3, 4 or 6.

Using these thermal functions with 16 adaptable constants, the minimisation of the mean error is performed with all final cycles for all four temperatures of one rubber compound. Start values are $A_k = 1$ and $B_k = 0$, with the conditions $A_k > 0$ and $B_k = 0$ fixed. After some iterations, all B_k are varied too, and after about 15 to 25 iterations the optimised solution is found. The mean error lies between 1 and 2% for all six compounds.

However a closer look at the results plotted for one temperature shows that the distribution of the deviations of calculated and measured hysteresis

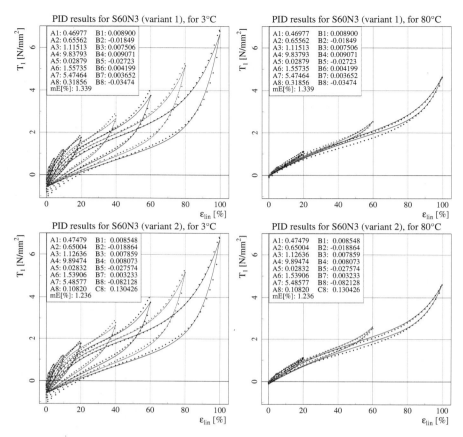

Fig. 4 MORPH adaptations with variants 1 and 2 for rubber compound S60N3 at 3 and 80°C.

curves are not acceptable for the highest temperatures. Figure 4 shows the result for the compound S60N3 at 3 and 80°C, the adaptation is performed by using the data of all four temperatures. The calculated hysteresis area at 80°C is too small compared to the measured data. The deviations at 3°C are in similar size, especially at low strains for cycles with large amplitudes, but here their effect is still acceptable due to the large hysteresis area of strong damping at this temperature.

One way to get a better fit for weakly damped curves is a higher weight for these curves at the expense of the others – but it is better to alter the thermal functions to get improved approximations. To set parameter p_6 temperature independent – fixed $B_6 = 0$ – shows only a negligible small change in mean error and error distribution when all other 15 constants are newly adapted. So in a second attempt, the parameter p_6 is fixed and an additional constant is spent for parameter p_8, which strongly influences the hysteresis height in

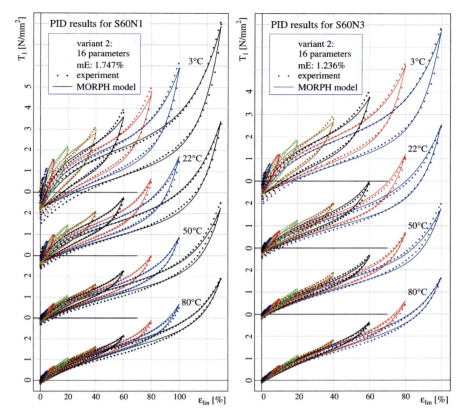

Fig. 5 MORPH adaptations for rubber compounds S60N1 and S60N3, shifted per temperature.

weakly damped cycles. This leads to the second variant for thermal effects in the MORPH model

$$\text{variant 2:} \quad p_k = A_k \cdot \exp(B_k \, \Delta\vartheta) \quad \text{for} \quad k = 1\ldots 5,\, 7,$$
$$p_6 = A_6, \quad (8)$$
$$p_8 = A_8 \cdot \exp(B_8 \, \Delta\vartheta) + C_8 \quad \text{with} \quad C_8 \geq 0.$$

In Figure 4 the improvements can be seen comparing plots of selected cycles for both variants, especially at 80°C. In variant 2 the hysteresis curves are fitted better. Here, for both variants the values of all A_k, B_k and C_8 are given.

This second variant is now applied to all six rubber compounds with good approximations for all measured temperatures. So this variant is considered to be applicable to technical rubber compounds. In Figures 5 and 6 the results

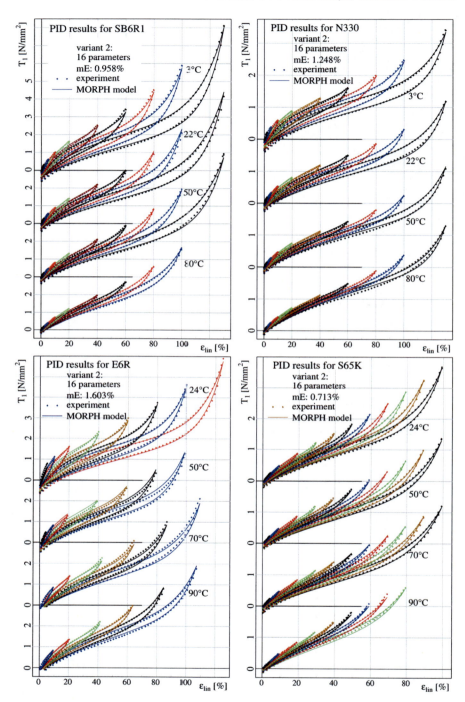

Fig. 6 MORPH adaptations for rubber compounds SB6R1, N330, E6R and S65K.

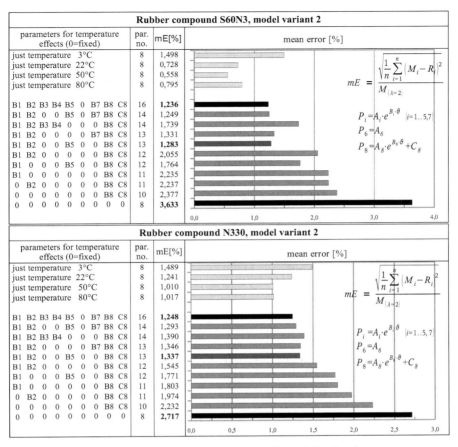

Fig. 7 MORPH adaptations with variant 2 and partially fixed parameters.

are shown for all six rubber compounds, the plots are shifted per temperature for better clarity.

At last the number of adaptable constants within variant 2 shall be reduced, if possible. All constants A_k are needed as well as B_8 and C_8. But the parameters B_1 to B_7 may be kept fixed to the value 0 so that the corresponding MORPH parameters p_1 to p_7 may be temperature independent. The task is to find combinations with temperature independent parameters p_k, which give good results for all six compounds and not only for one of them. Fixing one B_k and adapting all others, mostly leads to a mean error just slightly above the optimum of variant 2 with 16 adapted constants. E.g. setting p_5 or p_7 temperature independent is well acceptable, but setting both parameters temperature independent will increase the mean error much stronger!

A selection of these results is shown in Figure 7, for two rubber compounds as example. The first four rows represent the optimisations for just one

Fig. 8 Mean errors for MORPH parameter adaptations for six rubber compounds: Standard MORPH model without any temperature effects, variants 1 and 2 with 16 parameters and variant 3 as subgroup of variant 2 with 13 parameters.

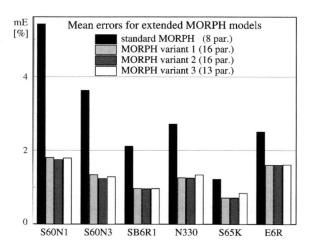

temperature with the standard MORPH model. The next row is the result of variant 2, using all final cycles at all measured temperatures of one compound. The mean error lies between the errors for each single specimen and includes expected variations for the four test pieces. This row with the black line is the optimum reference for all subversions with less than 16 constants whereas the last row represents the worst subversion – no temperature effect at all. All subversions with mean errors under 1% or near the optimum value and far away from the worst case can be considered as suitable versions, if this applies to all six compounds.

The best subversion for all six rubber compounds is variant 3 with 13 parameters in row 9 with the dark gray line.

$$\text{variant 3: } p_k = A_k \cdot \exp(B_k \, \Delta\vartheta) \quad \text{for} \quad k = 1, 2, 5,$$
$$p_k = A_k \quad \text{for} \quad k = 3, 4, 6, 7, \qquad (9)$$
$$p_8 = A_8 \cdot \exp(B_8 \, \Delta\vartheta) + C_8 \quad \text{with} \quad C_8 \geq 0.$$

In Figure 8 the mean errors for all six rubber compounds are collected, the first column for the standard MORPH model without any temperature effect, but adaptation of the eight parameters for all final cycles of one compound. The following 3 columns show the mean errors for the three discussed variants. Variant 3 with 8 MORPH constants at room temperature and 5 constants for temperature dependency fits the experimental data very well.

For several final cycles Figure 9 shows the effect of referring the discrete errors to a stress value representing the hysteresis height in the middle of the cycle with 100% maximum strain, different for each temperature. The weakly damped cycles are fitted better at the expense of the stronger damped cycles. So the hysteresis areas are approximated more adequately, representing damping and temperature rise.

Thermal Effects and Dissipation in a Model of Rubber Phenomenology 109

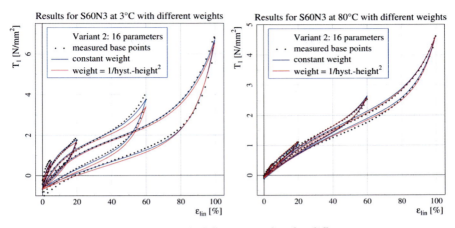

Fig. 9 MORPH adaptations with different weights for different temperatures.

4 Reversible Energy and Irreversible Dissipation

The MORPH rubber model only deals with stresses and strains. The stresses are built as a sum of hydrostatic stresses and deviatoric Neo-Hooke stresses and so called additional deviatoric stresses for nonlinear elastic and inelastic behaviour, see equation set (3). The first two are considered to be reversible, only the last term produces dissipation and reversible internal energy. Without external heat conduction or heat transfer, the difference between the internal power per volume and the reversible energy density rate will be irreversible dissipation due to internal damping, which has to be nonnegative with respect to the second law of thermodynamics. This dissipation will lead to a temperature rise, depending on rubber density and specific heat capacity due to the first law of thermodynamics.

$$
\begin{aligned}
& p_{\mathrm{in}}=\dot{w}_{\mathrm{in}}=\dot{\psi}_{\mathrm{el}}+p_{\mathrm{irr}}, \\
& p_{\mathrm{in}} \geq \dot{\psi}_{\mathrm{el}} \quad \Longleftrightarrow \quad p_{\mathrm{irr}}=\dot{w}_{\mathrm{irr}} \geq 0, \\
& \Delta \vartheta=w_{\mathrm{irr}} /(c\,\rho) \quad \text{for} \quad \underline{q}=\underline{0}.
\end{aligned}
\tag{10}
$$

Here, w_{irr} is the dissipative irreversible work, p_{irr} the dissipative irreversible power, w_{in} the internal work, p_{in} the internal power, ψ_{el} the elastic energy density, all related to volume of reference configuration. If the work of the additional stresses is assumed to be purely irreversible, then the second law of thermodynamics is not fulfilled in a short period directly after the reversal points at the upper and lower displacement limits because there the dissipation rate becomes negative till the hysteresis curve cuts the line of Neo-Hooke stress, see Figure 10. Since the additional stresses provide dissipative and nonlinear elastic behaviour, it is necessary to define an additional function for the nonlinear elastic parts of the MORPH model.

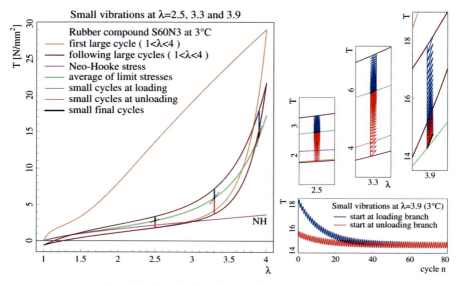

Fig. 10 Small vibrations at large deformations.

For all following simulations the extended MORPH model variant 2 is applied, where the MORPH parameters are thermal functions. So the additional elastic function may depend on some new constants and the eight MORPH parameters without further temperature effects. Stress-strain plots and dissipative work plotted versus strain are independent of strain rates, but dissipation rate and work plotted versus time depend on strain rate progress. Simulations are calculated with constant or sinusoidal deformation rates. In all plots, large cycles have constant deformation rates and small vibrations are performed at sinusoidal deformation rates. Since the MORPH routines in our program are based on the Lagrangian formulation, the elastic energy density will be formulated using $\underline{C}$, its invariant $C_T \equiv b_T$ and the history invariant $C_T^H \equiv b_T^H$.

4.1 Large and Small Tensional Cycles

In Figure 10, 1st PK stresses T are plotted versus stretch ratio λ for five large cycles $1 < \lambda < 4$ followed by 300 small cycles with amplitudes of 0.5% at a mean value of $\lambda = 2.5$, 3.3 or 3.9 respectively, starting at the loading branch of the last large cycle or at its unloading branch. Here the results of six different simulations are shown in one diagram. The final cycles with small amplitudes are approached after many cycles, like in the plots at the right of Figure 10. They are specially drawn as thick black lines and are identical for starting at loading or unloading branch. The lines for Neo-Hooke stress and for the average of loading and unloading limit stresses of the final large cycle are added in magenta and green lines. All three final small cycles touch

Thermal Effects and Dissipation in a Model of Rubber Phenomenology   111

or cut the line of average limit stresses, so these stresses should be a better choice for the elastic stresses than the Neo-Hooke stresses.

Despite of the rubber damping seen in the hysteresis curve for large amplitudes, the final small cycles show just vanishing damping and are nearly elastic. But their gradients – the black dotted lines – are larger than the gradient of the green line! So even this green line as elastic stress would still violate the second law of thermodynamics, but the error would be much smaller.

The main problem is that the green line is calculated as average of the limit stresses for loading and unloading, they depend on deformation rate tensors and cannot be used for an elastic potential. This applies especially to nonperiodic deformations.

It will be difficult to find a formulation for the reversible energy density which fits the demands for small and large cycles. If the formulation depends only on $\underline{C}$, C_T, C_T^H and the temperature dependent MORPH parameters, then e.g. at $\bar{\lambda} = 3.3$ this function would have the same value for loading and unloading branches of large and small cycles. That cannot meet the demand for nonnegative dissipation rates in all situations.

4.2 Derivation of an Energy Density for Additional Stresses

Hence Besdo suggested to take into account the additional stresses themselves. They are continuously differentiable and take the same values after complete periodic cycles (except the first one). So a scalar function for the reversible internal energy density is required, which might depend on $\underline{C}$, C_T, C_T^H and $p_k(\vartheta)$ and the additional stresses $\underline{\underline{\sigma}}^A$, $\underline{\underline{T}}^A$ or $\underline{\underline{\tilde{T}}}^A$. Here, simulations for large uniaxial tension and compression are compared to find the best fit for using the same kind of invariant for one of these three stress tensors. The principal approach for the energy density per volume of reference configuration is

$$\psi_{\text{el}} = \psi^h + \psi^{NH} + \psi^A + \psi_{\text{therm}}(\vartheta) \quad \text{with}$$

$$\psi^h = \frac{K}{2}(J - J_\vartheta)^2$$

$$\psi^{NH} = \alpha_e \left[\left(\frac{J_\vartheta}{J}\right)^{2/3} \underline{\underline{C}} \cdot \cdot \underline{\underline{1}} - 3 \right] \quad (11)$$

$$\psi^A = f(g_i(\ldots)).$$

For g_i three invariants of deviatoric stresses are considered, see Equations (2), and a normalised form of $\underline{\underline{C}}$ with $\det(\underline{\underline{C}}) = 1$:

$$g_0 = \underline{\underline{\sigma}}_D^A \cdot \cdot \underline{\underline{\sigma}}_D^A = (\underline{\underline{\tilde{T}}}_D^A \cdot \underline{\underline{C}}) \cdot \cdot (\underline{\underline{\tilde{T}}}_D^A \cdot \underline{\underline{C}})$$

$$g_1 = \underline{\underline{T}}_D^A \cdot \cdot (\underline{\underline{T}}_D^A)^T = (\underline{\underline{\tilde{T}}}_D^A \cdot \underline{\underline{C}}) \cdot \cdot \underline{\underline{\tilde{T}}}_D^A \quad (12)$$

$$g_2 = \underline{\underline{\tilde{T}}}_D^A \cdot \cdot \underline{\underline{\tilde{T}}}_D^A$$

If only uniaxial tension is considered, one could use any of these three functions and compensate the differences by scalar functions of C_T, but considering large tension and large compression it is not possible to compensate the differences, since C_T increases with larger tension and larger compression.

In the upper plots of Figure 11 all three stresses are plotted versus λ for two temperatures and two different simulations: uniaxial tension cycles with increasing upper limits $\lambda_a = 1.5, 2.0, 3.0$ and uniaxial compression cycles with decreasing lower limits $\lambda_a = 0.75, 0.5, 0.35$, both simulations starting with history invariant $C_T^H = 0$. The absolute Cauchy stresses are very small for compression, the absolute 2nd PK stresses very large compared to tension. These effects will become smaller due to the deviatoric stress tensors, see Equation (2). Simulations show that applying function g_0 with optimised functions of C_T and C_T^H, the elastic potential is too small for compression or too large for tension. When applying g_2 the elastic potential is too large for compression or too small for tension. The best fit can be achieved with function g_1, so from now on it will be used.

In the bottom plots of Figure 11 the dissipation rate of final cycles is plotted versus λ for the above described simulations. The black and blue dashed lines are calculated without elastic part of the additional stresses and show negative values after all reversal points, especially at large deformations, this violates the second law of thermodynamics. The red and brown lines show the improvement of an elastic function for the additional stresses: the dissipation rates increase at the beginning of unloading branches, but decrease near the reversal points of loading branches. If the factor in the formula for ψ^A is too small, the brown lines for unloading still have some negative values, if the factor is too large, the red lines for loading will get negative values just before reaching the maximum load. For large compression the loading branch approaches 0, so the factor in ψ^A may not be larger. For large tension the unloading branch approaches 0 as well, so this factor may not be smaller. Hence it is the best fit with a formula explained later, see Equation (13).

In the middle of Figure 11 the dissipated work is shown with and without ψ^A, these curves should increase with time everywhere. The black solid line with ψ^A is strictly increasing, whereas the dashed blue line without ψ^A is decreasing especially after all reversal points. Both lines have the same values at $\lambda = 1$, so the dissipated work per cycle is not changed by ψ^A. Additionally, the curve for ψ^A itself is plotted.

Applying g_1 for ψ^A an appropriate function $f(g_1)$ must be found to fit small and large amplitudes with a starting history invariant of $C_T^H = 0$ or $C_T^H > C_{T\max}$. Setting ψ^A proportional to g_1 will lead to elastic stresses, which are either too small in small cycles or too large in large cycles. One attempt to avoid the strong increase in large tension is a logarithmic function of g_1, compensating the exponential function in the limit stresses. But it turns out that the elastic part still increases too strong for large amplitudes.

Since the elastic part of the additional stresses must lie in the range of these additional stresses, another attempt applies the square root of g_1, a

Thermal Effects and Dissipation in a Model of Rubber Phenomenology

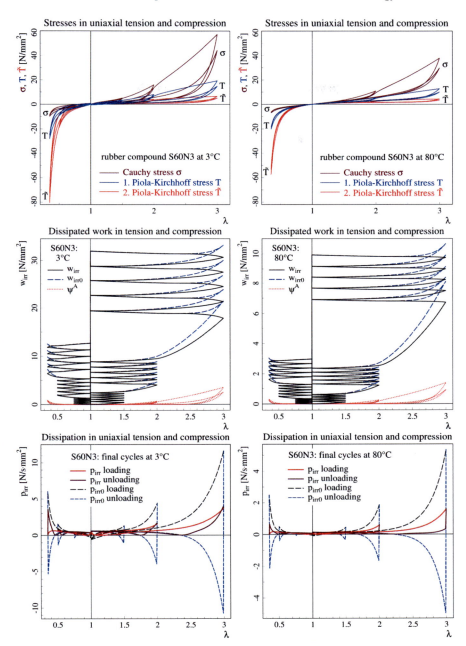

Fig. 11 Stresses and dissipation for simulations of uniaxial tension cycles $1 < \lambda < 3$ and uniaxial compression cycles $0.35 < \lambda < 1$, both groups starting with $C_T^H = 0$. $w_{\text{irr}0}, p_{\text{irr}0}$ are calculated without ψ^A, whereas $w_{\text{irr}}, p_{\text{irr}}$ are calculated with optimised function ψ^A.

kind of von Mises stress invariant. For large deformations this leads to good results, but for small values of g_1 the derivative $d\psi^A/dg_1 \to \infty$. Since the value $g_1 = 0$ is reached whenever the hysteresis curve crosses the line of Neo-Hooke stress, the formulation must be extended for small g_1 and for small cycles

$$f = a \cdot (\sqrt{g_1 + c^2} - c) \quad \text{with} \quad \lim_{g_1 \to 0}\left(\frac{df}{dg_1}\right) = \frac{a}{2c}, \quad \lim_{g_1 \to 0} f = \frac{a}{2c} \cdot g_1.$$

The constant c can be used to fit the additional elastic energy density for small cycles, where the effective elasticity is dominated by the exponential approach of the additional stress to the corresponding limit stress.

The strongest violations of the second law of thermodynamics occur just after the reversal points at large deformations λ_a where the invariant C_T is near to C_T^H. Here the additional stresses are dominated by the exponential term in the limit stresses, compare Equation (3). Setting $\lambda = \lambda_a - \delta$ with large λ_a and small δ, one can approximate for the additional elastic stress in uniaxial tension

$$\sigma_{el}^A \approx \sigma_0 \exp\left[p_7 \frac{\overset{*}{b_1}(\lambda^2 - \lambda^{-1})}{\overset{*}{b_T}(\lambda_a^2 - \lambda_a^{-1})}\right].$$

This can be developed to the additional elastic 1st PK stress

$$T_{el}^A = \frac{1}{\lambda}\sigma_{el}^A = \frac{d\psi^A}{d\lambda} = -\frac{d\psi^A}{d\delta} \approx \frac{\sigma_0}{\lambda_a}\exp\left[p_7(1 - \frac{2\delta}{\lambda_a})\right].$$

Integration yields approximately

$$\psi^A \approx \frac{\sigma_0}{2p_7}\exp\left[p_7(1 - \frac{2\delta}{\lambda_a})\right] \quad \Longrightarrow \quad \psi^A \sim \frac{a\,g_1}{2p_7}.$$

Without p_7 in the denominator, simulations require a temperature dependent constant a, increasing for rubber compound S65K, slightly increasing for E6R and SB6R1, and decreasing for S60N1, S60N3 and N330. That corresponds just to $1/p_7$, where the constant B_7 is positive for S60N1, S60N3 and N330, and negative for E6R, S65K and SB6R1. So with p_7 in the denominator for ψ^A, the constant a may be set temperature independent.

At last, the elastic energy density is expanded with a term related to the ratio C_T/C_T^H. If this ratio gets small, the exponential term in the additional stresses has only small influence compared to the constant term with p_8. So another scalar factor is introduced to take care for $C_T < C_T^H$ resulting in the final equation

$$\psi^A = \frac{a(b + C_T^H)}{2p_7(b + C_T)}\left[\sqrt{(\underline{\tilde{T}}_D^A \cdot \underline{C}) \cdot \underline{\tilde{T}}_D^A + c^2} - c\right] \tag{13}$$

Simulating uniaxial tension and compression or simple shear, the three new constants a, b and c can be adapted to fit the requirement of nonnegative dissipation rates in most cases and to violate the second law of thermodynamics only with negligible small values in some special situations. For all six rubber compounds, these new constants can be set to

$$a = 0.9 \cdots 1.0, \quad b = 1.2 \cdots 1.6, \quad c = 1.5 \cdots 3. \tag{14}$$

These numbers come out to be independent from temperature.

4.3 Simulations with New Energy Density

All plots for stresses, dissipation work or power are calculated with the extended MORPH model variant 2 and an additional elastic energy density according to Equations (13) and (14). Large cycles are simulated at constant deformation rates and small vibrations at sinusoidal deformations. Simulations are performed with one set of constants a, b, c for each rubber compound. Here the results for rubber compound S60N3 are presented, applying the set of constants $a = 0.9$, $b = c = 1.6$ in all simulations.

The first results are plotted in Figure 11 showing nonnegative dissipation rates for all cycles. In Figure 12 the dissipation in final cycles is shown for the same deformation cycles as in Figure 11, but here the starting history invariant is set to $C_T^H = 10$ due to any large previous deformation. The dissipation is smaller and the exponential term in the limit stresses has only small influence. A well adapted constant b yields nonnegative dissipation rates in all cycles – also in the first cycles which are not plotted for better clarity.

In Figure 13, 1st PK stresses and dissipation rates are shown for rubber compound N330 in uniaxial tension and compression like in Figure 11.

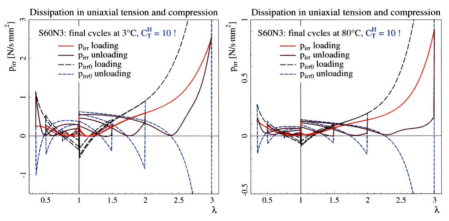

Fig. 12 Dissipation rates for uniaxial tension and compression with $C_T^H = 10$.

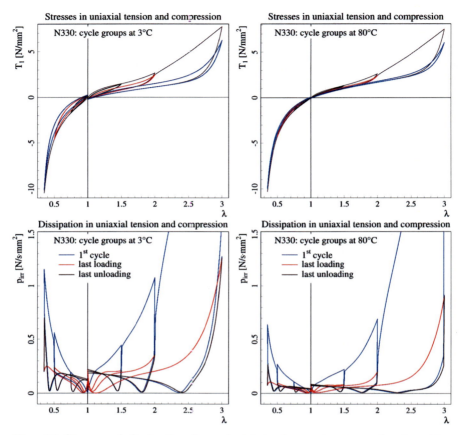

Fig. 13 Stress and dissipation for uniaxial tension and compression in rubber compound N330.

This time, the bottom plots show only the dissipation rates simulated with the energy density including optimised ψ^A. They are strictly nonnegative as postulated.

In Figure 14 stress T_{21} and dissipation are shown for simple shear with amplitudes $F_{12} \leq 1$ for a temperature of 3°C with high damping and for 80°C with weak damping. The hysteresis curves in the stress-strain plots differ strongly for these temperatures, but the plots for dissipation rate look very similar except for the scale. The dissipation rate is nonnegative even for cycles at $F_{12} < 0.5$ after large cycles with $F_{12} < 1$. So the dissipation work is generally increasing.

Figure 15 shows dissipation work and power for small sinusoidal vibrations of uniaxial tension, again for two temperatures with strong and weak damping. The strains vary from 0 to 1, 2 and 5%. Here the dissipation rate takes the value 0 at all reversal points, because the deformation velocity is 0. But

Thermal Effects and Dissipation in a Model of Rubber Phenomenology 117

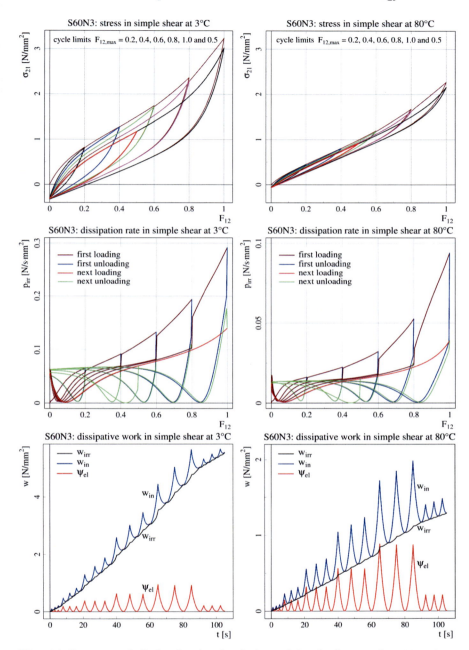

Fig. 14 Stresses and dissipation for simulations of simple shear cycles at two temperatures. w_{irr} is the dissipative work, w_{in} the internal work and ψ_{el} the elastic energy density, all related to reference volume.

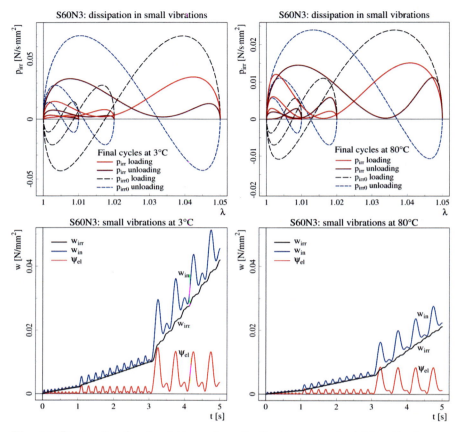

Fig. 15 Dissipation for simulations of small vibrations in uniaxial tension $0 < \epsilon < \epsilon_a$ with $\epsilon_a = 1$, 2 and 5%.

in all cases the dissipation rates for loading or unloading stay nonnegative. The dashed lines show simulations without ψ^A with large negative sections.

Finally, the introducing simulation for this section in Figure 10 is considered again. For instance, Figure 16 shows the dissipation work versus time for five large cycles $1 < \lambda < 4$ at constant deformation rate, followed by small sinusoidal vibrations with amplitudes of 0.5% strain at a large deformation $\lambda = 3.9$. The large cycles show monotonically increasing dissipated work. The small vibrations are virtually elastic, and in this case insignificant small negative gradients occur, since the exact peak to peak gradient depends on the distance between the limit stresses for loading and unloading and on the value of β in the differential equation (3). These stresses are functions of the deformation rate tensor and not suitable for elastic potentials. The negative dissipation rates are so small that they cannot be seen in the left-hand plot

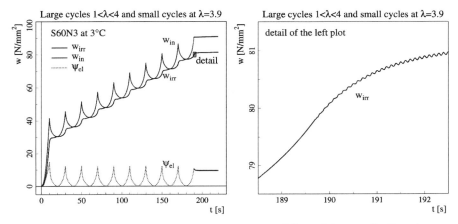

Fig. 16 Dissipation for large cycles $1 < \lambda < 4$ followed by small vibrations at $\lambda = 3.9 \pm 0.005$. w_{irr} is the dissipative work, w_{in} the internal work and ψ_{el} the elastic energy density, all related to reference volume.

of Figure 16, but the zoomed detail in the right-hand plot shows very small oscillations in the irreversible work.

4.4 Discussion

Neglecting heat conduction and transfer, differences in dissipative work within one time interval will lead to a temperature rise of approximately $\Delta\vartheta = \Delta w_{\mathrm{irr}}/(c\,\rho)$. Setting $c = 1.9$ kJ/kg·K and $\rho = 1.1$ kg/dm^3 the temperature rise for large cycles in Figure 16 can be estimated: $\Delta w_{\mathrm{irr}} = 5.84$ N/mm^2 $\Longrightarrow$ $\Delta\vartheta_{\mathrm{per}} = 2.8°$C. For small cycles after the first hundred ones the temperature rise per cycle is $\Delta\vartheta_{\mathrm{per}} = 0.0006°$C. During such a cycle heating and improper cooling occurs with $\Delta\vartheta_{\mathrm{heating}} = +0.0241°$C and $\Delta\vartheta_{\mathrm{cooling}} = -0.0235°$C. Since these small cycles are nearly elastic, it takes about 1700 cycles for 1°C heating. For finite element calculations the insignificant small heating and cooling sections within each cycle of 0.024°C are negligible.

Principally negative dissipation rates cannot occur in reality. Whenever the extended MORPH rubber model simulates small negative sections, it is possible to avoid this effect by an additional reserve energy density ψ^{res} due to differences between physical behaviour and phenomenological based simulations. The usage of such a reserve energy with a base value ψ_0^{res} may be defined by the following set of equations:

$$\begin{aligned}
&\text{if } \bar{p}_{\mathrm{irr}} < 0 &&\Longrightarrow && \dot{\psi}^{\mathrm{res}} = \bar{p}_{\mathrm{irr}}, && p_{\mathrm{irr}} = 0,\\
&\text{if } \bar{p}_{\mathrm{irr}} > 0 \text{ and } \psi^{\mathrm{res}} < \psi_0^{\mathrm{res}} &&\Longrightarrow && \dot{\psi}^{\mathrm{res}} = \bar{p}_{\mathrm{irr}}, && p_{\mathrm{irr}} = 0, && (15)\\
&\text{if } \bar{p}_{\mathrm{irr}} > 0 \text{ and } \psi^{\mathrm{res}} \geq \psi_0^{\mathrm{res}} &&\Longrightarrow && \psi^{\mathrm{res}} = \psi_0^{\mathrm{res}}, && p_{\mathrm{irr}} = \bar{p}_{\mathrm{irr}}.
\end{aligned}$$

Fig. 17 Effect of reserve energy density to small vibrations at large deformation, visible just with large magnification: The dissipation work oscillates without reserve energy ($\bar{w}_{irr}$) and increases monotonically, when reserve energy is applied (w_{irr}).

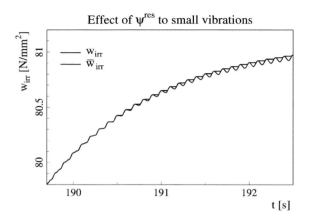

Here $\bar{p}_{irr}$ denotes the so far calculated dissipation without reserve energy density, while p_{irr} stands for the final dissipation per reference volume, which will never become negative.

Figure 17 shows the effect of this reserve energy density in a strongly magnified section of Figure 16. The improper oscillations of the dissipative work without reserve energy are smoothed, and now with reserve energy the dissipation rate is strictly nonnegative. During each cycle the reserve energy density will come up to the base value ψ_0^{res} and then the dissipation work will be increased. In sections with dissipation rate 0 the reserve energy is applied and filled up again, and for positive dissipation rates the reserve energy remains constant.

But considering these small violations of nonnegative dissipation rates, it does not seem to be necessary to implement this reserve energy density in finite element routines.

5 Conclusion

The MORPH rubber model is extended to temperature dependent stress-strain curves by setting seven (variant 2) or only four (variant 3) of the eight original parameters as temperature dependent. This leads to good results for all experimental data provided by the DIK in Hannover. The introduction of an elastic energy density for the additional stresses of the MORPH model yields nonnegative dissipation rates for most deformation processes. Only insignificant negative sections may occur in small, nearly elastic vibrations in a strongly deformed configuration. For special cases, a reserve energy density may be applied to enforce strictly nonnegative dissipation rates.

References

1. Besdo, D., Ihlemann, J.: The effect of softening phenomena in filled rubber during inhomogeneous loading. In: Constitutive Models for Rubber II, Balkema, Rotterdam, pp. 137–147 (2001)
2. Besdo, D., Ihlemann, J.: Wechsel zwischen Eulerscher und Lagrangescher Formulierung mit einem speziellen Übertragungsoperator. Proc. Appl. Math. Mech. 1, 143–144 (2001)
3. Besdo, D., Ihlemann, J.: Effiziente Parameteridentifikation bei Stoffgesetzen für inelastische technische Gummimaterialien. Proc. Appl. Math. Mech. 3, 324–325 (2003)
4. Besdo, D., Hohl, C., Ihlemann, J.: ABAQUS implementation and simulation results of the MORPH constitutive model. In: Constitutive Models for Rubber IV, Balkema, Leiden, pp. 223–228 (2005)
5. Besdo, D., Hohl, C.: Einfluss inelastischer Effekte auf die Reibung in zyklischen Prozessen, Zwischenbericht zum DFG-Projekt Forschergruppe FOR 492/1, Teilprojekt TP2a, Hannover (2005)
6. Hohl, C.: Anwendung der Finite-Elemente-Methode zur Parameteridentifikation und Bauteilsimulation bei Elastomeren mit Mullins-Effekt. In: Fortschr.-Ber. VDI Reihe, vol. 18(310). VDI Verlag, Düsseldorf (2007)
7. Ihlemann, J.: Kontinuumsmechanische Nachbildung hochbelasteter Gummiwerkstoffe. In: Fortschr.-Ber. VDI Reihe, vol. 18 (288). VDI Verlag, Düsseldorf (2003)
8. Lorenz, H., Klüppel, M.: Micromechanics of internal friction of filler reinforced elastomers. In: Besdo, D., et al. (eds.) Elastomer Friction – Theory, Experiment and Simulation. LNACM, vol. 51, pp. 27–52. Springer, Heidelberg (2010)

Finite Element Techniques for Rolling Rubber Wheels

U. Nackenhorst, M. Ziefle, and A. Suwannachit

Abstract. Arbitrary Lagrangian Eulerian (ALE) methods provide a well established basis for the numerical analysis of rolling contact problems. Whereas the theoretical framework is well developed for elastic constitutive behavior, special measures are necessary for the computation of dissipative effects like inelastic properties and friction because the path of material points is not traced inherently. In this presentation a fractional step approach is suggested for the integration of the evolution equations for internal variables. A Time-Discontinuous Galerkin (TDG) method is introduced for the numerical solution of the related advection equations. Furthermore, a mathematically sound approach for the treatment of frictional rolling within the ALE description is suggested. By this novel and fully implicit algorithm the slip velocities are integrated along their path-lines. For dissipative effects due to both, inelastic behavior and friction, physical reliable results will be demonstrated as well as the computability of large scaled finite element tire-models.

1 Introduction

The analysis of rolling contact problems is of broad industrial importance, e.g. for the optimization of roller bearings, wheel-rail systems or car tires. Early analytical approaches, for example proposed by Carter [7], are based on half-space assumptions and therefore, limited to linear investigations. Computational techniques for rolling contact problems are under development since more than two decades, besides many other contribution we like to cite the pioneering work from Oden and Lin [26]. Already in this early work a suitable relative kinematic framework has been introduced and an engineering

---

U. Nackenhorst · M. Ziefle · A. Suwannachit
Institut für Mechanik und Numerische Mechanik, Leibniz Universität Hannover,
Appelstr. 9A, 30167 Hannover, Germany
e-mail: {nackenhorst,ziefle,suwannac}@ibnm.uni-hannover.de

approach for viscoelastic material behavior has been suggested. A continuum mechanics generalization of the relative kinematic description in an Arbitrary Lagrangian Eulerian (ALE) description has been outlined in [24].

The advantages of the ALE approach for rolling contact analysis in a finite element framework are summarized as follows:

- Stationary rolling is described independent of time, because temporal derivatives are replaced by spatial gradients. Therefore, stationary rolling contact problems are computed like quasi static analysis.
- The spatially fixed reference configuration allows for local mesh refinement, for example for a detailed analysis of the contact reactions.

However, in contrast to the rapid development of ALE methods in other fields of engineering application, like fluid-structure interaction or metal forming processes for example, compare [13], only few improvements have been made in rolling contact mechanics so far. One example is the treatment of inelastic material properties, where traditional engineering approaches are used to integrate the history variables along predefined concentric rings of integration points, see [11, 25, 26] for example. This approach seems to be established in the rolling contact community, though there are obvious problems with unstructured meshes and a sound mathematical basis is missing. A first step in this direction has been presented by Le Tallec and Rahier [23] who applied methods established in computational fluid mechanics for the continuous transport of viscoelastic history variables within the spatially fixed ALE mesh. A finite difference upwind-scheme for the solution of the underlying advection problem has been suggested. A further problem arises from the treatment of tangential contact conditions which is discussed controversially in literature, compare [18, 24, 33]. Because the material history of particles moving through the finite contact patch is not determined directly within the spatially fixed finite element mesh of rolling bodies, additional effort has to be spent to enforce stick conditions and related friction rules.

This contribution targets on both problems, basically investigated in [37]. In the first part a mathematical consistent theory for the solution of the advection for inelastic constitutive properties is described. For the numerical treatment of the evolution equations for arbitrary inelastic internal variables the so called fractional step strategy suggested by Benson [4] has been chosen. This enables for a local solution using well established and efficient numerical schemes, e.g. radial return mapping for the evolution of plastic strain. In a second step the transport of the internal variables corresponding to the material particles motion within the spatially fixed finite element mesh is tackled be the numerical solution of the corresponding hyperbolic advection equations. The advantage of Time-Discontinuous Galerkin (TDG) methods above more traditional finite difference and finite volume methods for advection dominated problems has been shown recently in [38], a related staggered scheme for the investigation of inelastic properties of rolling wheels will be introduced in Section 4.

The general idea of tracing material path lines is adapted for the treatment of the frictional contact of rolling wheels. In contrast to the prior problem of history dependent material properties here only a small domain of the overall finite element model is affected, which motivates for a fully implicit solution. The general idea for a mixed interpolation of both, the nodal displacements and the relative material motion of the contacting particles has been outlined in [39] and is summarized in Section 5.

The physical consistency and numerical reliability is demonstrated by three-dimensional computations for a solid rubber wheel, this so called Grosch wheel has been chosen as reference subject within the Research Unit 492. In addition, studies on detailed discretized tire models will be shown, by which the industrial applicability of the proposed approach is emphasized. The potential of the proposed computational mechanics techniques for a new insight into the physical mechanism of rolling contact phenomena is shown as well as the industrial applicability is demonstrated.

2   Relative Kinematic Framework for Rolling Contact

For the efficient numerical solution of rolling contact problems a suitable relative kinematic framework is needed, for which a special kind of Arbitrary Lagrangian Eulerian (ALE) method has been established when finite deformations have been taken into account. Here only the general idea is outlined, for more theoretical details it is referred to [24]. As depicted in Figure 1, the motion is decomposed into a pure rigid body motion described in Eulerian coordinates, leading to an intermediate reference configuration, denoted $\chi(B)$, and relative to this reference configuration the deformation is described by the mapping $\hat{\Phi}$ in Lagrangian coordinates. With the kinematic assumption, that the wheel is spinning around a spatially fixed axis while the road is moving with constant speed, the rigid body motion is described by a pure rotation tensor $\boldsymbol{R}$, which leads to a multiplicative split of the material deformation gradient,

$$\boldsymbol{F} = \hat{\boldsymbol{F}} \cdot \boldsymbol{R}. \tag{1}$$

Like in Eulerian mechanics the velocity of a material point splits into a relative derivative and a convective part,

$$\boldsymbol{v} = \left.\frac{\mathrm{d}\boldsymbol{\varphi}}{\mathrm{d}t}\right|_X = \left.\frac{\partial \boldsymbol{\varphi}}{\partial t}\right|_\chi + \mathrm{Grad}\,\boldsymbol{\varphi} \cdot \boldsymbol{w}, \tag{2}$$

while the gradient is computed with respect to the reference configuration and $\boldsymbol{w}$ describes the guiding velocity due to the rigid body motion. The variable $\boldsymbol{\varphi}$ describes the spatial location which is actually passed by an arbitrary material point. It is mentionable that for stationary motion the relative velocity vanishes, which means, that the material time derivative is replaced by spatial gradients.

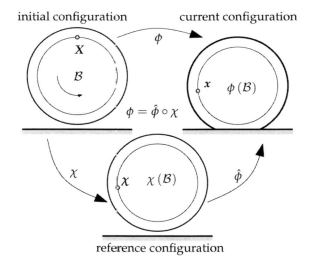

Fig. 1 ALE description of rolling wheels.

The balance of momentum is written with respect to the reference configuration, the corresponding weak formulation is referred to as

$$\int_{\chi(B)} \hat{\varrho}\,\dot{\boldsymbol{v}} \cdot \boldsymbol{\eta}\, d\hat{V} = -\int_{\chi(B)} \hat{\boldsymbol{P}} \cdot \cdot \operatorname{Grad} \boldsymbol{\eta}\, d\hat{V} + \int_{\chi(B)} \hat{\varrho}\,\boldsymbol{b} \cdot \boldsymbol{\eta}\, d\hat{V} + \int_{\partial_t \chi(B)} \boldsymbol{t} \cdot \boldsymbol{\eta}\, d\hat{A}, \tag{3}$$

with the First Piola–Kirchhoff representation of the stress tensor $\hat{\boldsymbol{P}}$ and the externally applied body loads $\boldsymbol{b}$ and surface traction $\boldsymbol{t}$, while $\boldsymbol{\eta}$ describes a test function in the Galerkin sense.

With the assumption of axi-symmetric material distribution the inertia term can be reformulated in a most symmetric form, see [24] for details. For stationary rotations it reads

$$\int_{\chi(B)} \hat{\varrho}\,\dot{\boldsymbol{v}} \cdot \delta\boldsymbol{\varphi}\, d\hat{V} = -\int_{\chi(B)} \hat{\varrho}\, (\operatorname{Grad} \boldsymbol{\varphi} \cdot \boldsymbol{w}) \cdot (\operatorname{Grad} \boldsymbol{\eta} \cdot \boldsymbol{w})\, d\hat{V} + \int_{\partial \chi(B)} \hat{\varrho}\, \delta\boldsymbol{\varphi} \cdot \boldsymbol{c}\, \boldsymbol{w} \cdot \hat{\boldsymbol{n}}\, d\hat{A}; \tag{4}$$

the first term at the right-hand side describes the virtual work due to centrifugal forces and by the second term the impulse flux over the boundary of the domain is taken into account. It is argued that this part vanishes at natural boundaries of a spinning wheel, because of the outward unit normal $\hat{\boldsymbol{n}}$ is always perpendicular to the guiding velocity $\boldsymbol{w}$. Nevertheless, this only holds for the continuous description; for the discretized problem additional effects have to be considered which has already been discussed in [24].

The linearized and discretized form of this nonlinear equation (3) results in the incremental finite element equation for the motion of steady-state spinning bodies,

$$[K - W][\Delta \varphi] = [f_e + f_i - f_\sigma]. \tag{5}$$

Herein K describes the tangential stiffness matrix and W is the ALE-inertia matrix. On the right-hand side f_e describes the externally applied forces, f_i are the equivalent nodal forces caused by inertia effects and f_σ describes the internal forces due to the divergence of the stress state.

For pure elastic response the proposed approach provides a framework for efficient numerical techniques, i.e. stationary rolling is described time independent, because time derivatives are replaced by spatial gradients. In addition, the spatially fixed reference mesh can be refined goal oriented for the contact analysis. However, additional effort has to be spent when inelastic constitutive behavior or frictional contact have to be taken into account, which will be outlined below.

3 Constitutive Modelling of Rubber

The mechanical behavior of rubber is phenomenologically characterized by incompressible and large elastic deformations, damage effects under initial loading (Mullins effect), hysteresis under quasi-static cyclic loading including remaining strain and viscoelastic effects under dynamic conditions. Many of these phenomena are explained by the molecular microstructure of vulcanized and particle filled polymers. This underlines the concept of entropy-elasticity within a thermodynamic consistent framework.

A variety of constitutive models have been analyzed in this project with respect to their approximation behavior and their computational efficiency, for details the reader is referred to [9, 37]. A universal thermodynamic consistent approach is summarized below.

A class of hyper-elastic constitutive models for rubber has been implemented and intensively tested in this project, for an overview of hyper-elastic models and related derivatives the reader is referred to [16]. For rubber usually incompressibility or slight incompressibility is assumed, for which a multiplicative split of deformation and an additive split of the free energy function

$$\psi = W(\bar{C}) + U(J) \tag{6}$$

is stated. Here we introduced the deviatoric part of the Right Cauchy–Green tensor $\bar{C} = J^{-2/3} C$, and $J = \det F$ is the Jacobian of the deformation gradient. A broad variety of suggestions for the volumetric part have been made in literature, for example

$$U(J) = \frac{\kappa}{2} \left(\frac{J^2 - 1}{2} - \ln J \right) \tag{7}$$

is a proper penalty function that fulfills the growth conditions, where κ has the meaning of a bulk modulus.

Hyper-elastic constitutive models for the large deformation response have been implemented, where optional Neo-Hooke, Mooney–Rivlin or the micromechanical motivated tube–model can be activated. The free energy function for the tube-model for example is written as

$$W(\bar{C}) = \frac{G_c}{2}\left[\frac{(1-\delta^2)(I_{\bar{C}}-3)}{1-\delta^2(I_{\bar{C}}-3)} + \ln\left(1-\delta^2(I_{\bar{C}}-3)\right)\right] + \frac{2G_e}{\beta^2}\sum_{A=1}^{3}\left(\bar{\lambda}_A^{-\beta}-1\right). \tag{8}$$

It is assumed as an additive decomposition of chemical bound and topological restraints, expressed by the shear modulus G_c and G_e respectively. $I_{\bar{C}}$ is the first invariant of the deviatoric part of the Right Cauchy–Green stretch tensor and $\bar{\lambda}_A$ represent the mean stretches. With the constitutive parameter δ the finite stretchability of the molecules chains is expressed, and with β the instantaneous relaxation behavior can be approximated. For more details, refer to [17].

Inelastic effects are described by introducing internal variables, for the basic concepts and computational aspects the reader is referred to [32] for example. Based on the suggestion in [31] a stress valued internal variable $\boldsymbol{\alpha}_V$ for the time dependent visco-elastic response is introduced. A free energy function is written as

$$W(\bar{C}, \boldsymbol{\alpha}_V) = W^\circ(\bar{C}) - \frac{1}{2}\bar{C}\cdot\cdot\boldsymbol{\alpha}_V + W_V(\boldsymbol{\alpha}_V) \tag{9}$$

with

$$W_V(\boldsymbol{\alpha}_V) = -(1-\gamma)\,W^\circ(\bar{C}) + \frac{1}{2}\bar{C}\cdot\cdot\boldsymbol{\alpha}_V. \tag{10}$$

In the limit to one-dimensional small deformation theory a simple three parameter viscoelastic rheological model can be associated to this approach, where $\boldsymbol{\alpha}_V$ is interpreted as time dependent stress observed at the dashpot. The coefficient γ then describes the ratio between the spring stiffness and the total Young's modulus.

From the second law of thermodynamics an evolution rule for the viscoelastic internal variables $\boldsymbol{\alpha}_V$ like

$$\dot{\boldsymbol{\alpha}}_V(t) + \frac{1}{\tau}\boldsymbol{\alpha}_V(t) = \frac{1-\gamma}{\tau}J^{\frac{2}{3}}\bar{\boldsymbol{S}}^\circ(t) \tag{11}$$

is obtained. Here

$$\bar{\boldsymbol{S}}^\circ(t) = J^{-\frac{2}{3}}\,\mathrm{DEV}\left[2\frac{\partial W^\circ(\bar{C}(t))}{\partial \bar{C}}\right] \tag{12}$$

represents the deviatoric part of the total Second Piola–Kirchhoff stress obtained from the derivative of the free energy function with respect to the

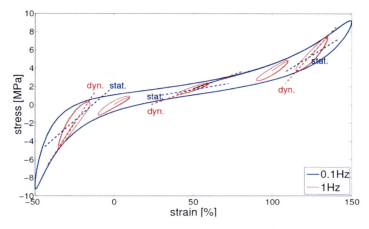

Fig. 2 Computed hysteresis loops for large amplitude/low frequency and small amplitude/high frequency response.

total Cauchy–Green stretch tensor; the operator is defined as DEV($\bullet$) = ($\bullet$) $- \frac{1}{3}(\bullet \cdot \cdot \boldsymbol{C})\boldsymbol{C}^{-1}$. The constitutive parameter τ is interpreted as relaxation time. Based on the concepts of linear visco-elasticity this approach is easily extended for a series of relaxation times for the approximation of viscoelastic behavior in a broad frequency domain. This on the other hand shows the limitations of this approach. Despite the fact, that the kinematics is written within a large deformation framework, for the viscoelastic response linear superposition behavior is assumed. Therefore, this modelling approach is referred to as finite linear visco-elasticity.

This approach enables for the representation of typical rubber behavior as shown in Figure 2. Depicted are the computational results for a quasi static (0.1 Hz) uniaxial cyclic loading with an amplitude of 100% strain surrounding a static pre-strain of 50%. The typical load-deflection curve as well as the hysteretic behavior is shown clearly. In addition the effect of dynamic stiffening is demonstrated in Figure 2, at certain strain levels the static load is kept fixed and the specimen is subjected to small displacement amplitudes with a frequency of 1 Hz. Here typical viscoelastic hysteresis ellipses are observed while the average slope increases in comparison with the tangents constructed at the quasi static response curve.

For particle reinforced rubber, like carbon black or silica reinforcement, the additional stress softening phenomenon has to be considered. Two different modeling concepts to simulate the so called Mullins effect have been implemented. The first approach follows a continuum damage mechanics model already presented in [31]. Optionally the pseudo-elastic modelling approach following Ogden and Roxburgh [27] is enabled. Applications of both damage approaches with Gao's elastic law were studied in [15]. The computational results show that in principle both models can describe the typical Mullins effect. However, these studies only deal with nonlinear hyperelastic behavior and damage effects, while rate dependent behavior is not considered. Therefore, the modeling of damage effects within the framework of viscoelasticity is studied and discussed in this work.

3.1 Continuum Mechanics Damage Model

In the following a brief outline of the continuum damage model introduced by Simo [31] is presented. According to an assumption that the bulk response is much larger than the deviatoric response in case of incompressible rubber materials, the damage mechanism is then restricted to the deviatoric part only. The deviatoric part of the free energy function is modified as

$$W(\bar{\boldsymbol{C}}, \boldsymbol{\alpha}_V, \alpha_D^c) = (1 - \alpha_D^c) W^\circ(\bar{\boldsymbol{C}}) - \frac{1}{2} \bar{\boldsymbol{C}} \cdot \cdot \boldsymbol{\alpha}_V + W_V(\boldsymbol{\alpha}_V), \qquad (13)$$

where a scalar valued internal variable $\alpha_D^c \in [0, 1]$ for the description of isotropic damage has been introduced. Here the notation $(\bullet)^\circ$ indicates the initial (undamaged) condition. The deviatoric part of the second Piola–Kirchhoff stress tensor can be consequently written in terms of damage parameter,

$$\bar{\boldsymbol{S}}(t) = (1 - \alpha_D^c)\,\bar{\boldsymbol{S}}^\circ(t) - J^{-\frac{2}{3}} \operatorname{DEV}\left[\boldsymbol{\alpha}_V(t)\right]. \qquad (14)$$

It is assumed that the damage initiation is described in dependency of a maximum equivalent strain measure defined as

$$\Xi^m(t) = \max_{s \in (-\infty, t]} \sqrt{2 W^\circ(\bar{\boldsymbol{C}}(s))}. \qquad (15)$$

Thus the damage criterion can be expressed by

$$f(\Xi(t)) = g(\Xi(t)) - \alpha_D^c(\Xi(t)) \leq 0. \qquad (16)$$

For $g(\Xi(t))$ any arbitrary function can be chosen which satisfies the conditions

$$g(\Xi^m) \in [0,1], \quad g(\Xi^m \to 0) = 0, \quad g(\Xi^m \to \infty) \geq 0. \qquad (17)$$

Following the principle of maximal internal dissipation the rate equation for the isotropic damage variable reads

$$\dot{\alpha}_D^c = \dot{\lambda}\, \frac{\partial f}{\partial Y}, \qquad (18)$$

with the Lagrange multiplier λ and the associated thermodynamical force Y. For a more detailed description of the continuum damage model the reader is referred to [31]. Within a pure Lagrangian description the evolution equations of the internal variables (11) and (18) can be integrated on element level by

Finite Element Techniques for Rolling Rubber Wheels

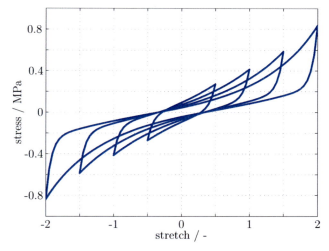

Fig. 3 Computed hysteresis in cyclic shear test of rubber (continuum damage mechanics model).

standard schemes which are well established in solid body mechanics, e.g. implicit Euler method.

A typical result is shown in Figure 3 where the hysteresis loops for a cyclic shear test with stepwise increased load amplitudes is depicted. In this graph only the stationary cycles are drawn. The softening behavior is clearly represented by the constitutive model, with increasing load amplitudes the stiffness decreases.

From the results of cyclic loading test it is concluded that damage is gradually accumulated and is presented already during the primary loading. This means the damage model influences both the loading path and the unloading or reloading paths. Therefore, the identification of constitutive parameters appears difficult since the contributions are mixed, so that the viscoelastic parameters and parameters of the damage model must be determined at the same time, see [15] for similar argumentations.

3.2 Pseudo-Elastic Damage Model

An alternative approach has been introduced by Ogden and Roxburgh [27] referred to as pseudo-elastic model. The material response is governed by different strain energy density functions for primary loading and unloading. On primary loading path the material response is controlled by a common energy density function of the viscoelastic model, e.g. Equation (9). It is assumed that no continuous damage takes place on this path. Just if the material is subjected to unloading, or reloaded again below the previous maximum driven strain, the deviatoric response will be controlled by a modified energy density function incorporating a scalar damage parameter $\alpha_D^{ps} \in [0,1]$.

The specific form of the modified energy density function suggested in [27] is written as

$$W(\bar{C}, \boldsymbol{\alpha}_V, \alpha_D^{ps}) = \alpha_D^{ps} W^\circ(\bar{C}) + \phi(\alpha_D^{ps}) - \frac{1}{2}\bar{C} \cdot \cdot \boldsymbol{\alpha}_V + W_V(\boldsymbol{\alpha}_V), \quad (19)$$

where $\phi(\alpha_D^{ps})$ is a damage function, which is related to the energy dissipated in a primary loading/unloading cycle.

The damage parameter α_D^{ps} is a switch set to be inactive and chosen to be 1 at any point on the virgin loading curve while the damage function satisfies $\phi(1) = 0$. In this case the pseudo-energy function is reduced to a common deviatoric part of the energy density function (9). On the unloading or reloading path below the previous maximum strain, α_D^{ps} varies in the range $0 < \alpha_D^{ps} < 1$.

To determine the corresponding damage function $\phi(\alpha_D^{ps})$ an additional equation

$$\frac{\partial W(\bar{C}, \boldsymbol{\alpha}_V, \alpha_D^{ps})}{\partial \alpha_D^{ps}} = 0 \quad (20)$$

arising from inclusion of the damage variable is introduced. With (19) one obtains

$$W^\circ(\bar{C}) = -\phi'(\alpha_D^{ps}), \quad (21)$$

which implicitly defines the damage parameter in terms of the deformation. With the assumption that $\alpha_D^{ps} = 1$ at any point on the primary loading path, the maximum energy density function is defined as

$$W_m = W^\circ(\bar{C}_m) = -\phi'(1). \quad (22)$$

Obviously, W_m increases along the primary loading path, until the unloading takes place and it will increase again if the further primary loading is initiated, so that

$$W_m = \max_{s \in (-\infty, t]} W^\circ[\bar{C}(s)]. \quad (23)$$

By using W_m as damage criterion, the deviatoric energy density function can be computed under two conditions as follows:

primary loading : if $W^\circ(\bar{C}) \geq W_m$

$$W(\bar{C}, \boldsymbol{\alpha}_V) = W^\circ(\bar{C}) - \frac{1}{2}\bar{C} \cdot \cdot \boldsymbol{\alpha}_V + W_V(\boldsymbol{\alpha}_V),$$

(24)

unloading or reloading: if $W^\circ(\bar{C}) < W_m$

$$W(\bar{C}, \boldsymbol{\alpha}_V, \alpha_D^{ps}) = \alpha_D^{ps} W^\circ(\bar{C}) + \phi(\alpha_D^{ps}) - \frac{1}{2}\bar{C} \cdot \cdot \boldsymbol{\alpha}_V + W_V(\boldsymbol{\alpha}_V).$$

Finite Element Techniques for Rolling Rubber Wheels 133

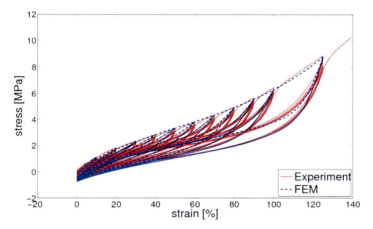

Fig. 4 Cyclic tensile test: comparison between experimental and computational results from the pseudo-elastic damage model.

In [27] the first derivative of the damage function

$$-\phi'(\alpha_D^{ps}) = m \operatorname{erf}^{-1}[r(\alpha_D^{ps} - 1)] + W_m, \qquad (25)$$

with a set of positive constitutive parameters r and m is proposed, where $\operatorname{erf}^{-1}(\bullet)$ is the inverse of the error function. Substituting (25) in (21), the damage variable can be finally expressed by

$$\alpha_D^{ps} = 1 - \frac{1}{r} \operatorname{erf}\left\{\frac{1}{m}[W_m - W^\circ(\bar{\boldsymbol{C}})]\right\}, \qquad (26)$$

which vanishes for virgin loading. The deviatoric part of the second Piola–Kirchhoff stress tensor

$$\bar{\boldsymbol{S}}(t) = \alpha_D^{ps} \, \bar{\boldsymbol{S}}^\circ(t) - J^{-2/3} \operatorname{DEV}[\boldsymbol{\alpha}_V(t)] \qquad (27)$$

is obtained from the first derivative of (19) with respect to $\boldsymbol{C}$.

A comparison between experimental data and computational results is depicted in Figure 4. In the experiment a rubber specimen is subjected to the uniaxial tension test under 10mm/min deformation rate and stretched in steps to 10, 20, 30, ..., 90, 100 and 120% strain. At each strain level five load cycles are performed until a stationary response is obtained. For the simulation similar loading conditions are applied to a single brick element. The computed material response is plotted as dotted lines. A quite good agreement between the experiment and simulation is obtained.

It should be noted that the pseudo-elastic model significantly simplifies the identification of constitutive parameters. Since damage is only activated during unloading and reloading, one can identify the viscoelastic parameters, i.e. relaxation time τ and stiffness ratio γ, and the parameters of the pseudo-elastic model (r and m) separately.

4 Treatment of Inelastic Behavior within the ALE Description of Rolling

The consideration of inelastic constitutive models requires tracking the history for each individual material particle. In a purely Lagrangian description this is straight forward and effective numerical schemes are well established, c.f. [32]. In the framework of the special ALE formulation an additional update procedure for the internal variables $\boldsymbol{\alpha}$ has to be provided for tracing the history of material particles which are moving in the spatially fixed mesh. In the ALE picture the evolution rules for the internal variables are affected by convective terms. The evolution rule of an internal variable $\boldsymbol{\alpha}$ is expressed by

$$\dot{\boldsymbol{\alpha}} = \left.\frac{\partial \boldsymbol{\alpha}}{\partial t}\right|_\chi + \operatorname{Grad} \boldsymbol{\alpha} \cdot \boldsymbol{w} = \mathcal{F}(\bar{\boldsymbol{C}}, \boldsymbol{\alpha}). \tag{28}$$

The guiding velocity $\boldsymbol{w}$ is computed by the angular velocity ω and the position vector of the particle in the reference configuration. The convective time derivative is then interpreted by the change of location of the corresponding material particle. In conclusion a time step is substituted by an angle increment $\Delta\gamma$ and the angular velocity,

$$\Delta t = \frac{\Delta \gamma}{\omega}. \tag{29}$$

Update procedures for the treatment of inelastic internal variables are discussed intensively in the scientific community also for other specific applications described by ALE methods, e.g. forming simulations of problems of fluid structure interaction. A survey of current ALE formulations and their applications is given in [10]. For the treatment of the convective terms two different solution strategies exist, the *split* or *fractional-step* strategy and the fully coupled or so called *unsplit* method, cf. [5].

4.1 The Fractional-Step Strategy

The *unsplit* method leads to a fully coupled systems of algebraic equations because the convective equations are integrated directly and are included into the weak form of the equation of motion, see [3] for example. In most applications the decoupled fractional-step strategy, introduced in [4, 5] appear to be favorable. Each time step is split into two phases: a pure Lagrange step and an Euler step, in which the material history variables are convected. The solution of the evolution equation (28) is split into two steps by solving sequentially the system

$$\left.\frac{\partial \boldsymbol{\alpha}}{\partial t}\right|_\chi = \mathcal{F}(\bar{\boldsymbol{C}}, \boldsymbol{\alpha}) \qquad \rightarrow \quad \text{Lagrange step,}$$
$$\left.\frac{\partial \boldsymbol{\alpha}}{\partial t}\right|_\chi + \text{Grad}\,\boldsymbol{\alpha} \cdot \boldsymbol{w} = \boldsymbol{0} \quad \rightarrow \quad \text{Euler step.}$$
(30)

In the prior Lagrange step the local evolution of the internal variables is evaluated depending on the current material history and the mechanical stress state by solving a parabolic partial differential equation. Algorithmic techniques for this task are well investigated, see [32] for example. An implicit integration scheme is easily implemented and computed at element level. In the subsequent Euler step the material history is transported through the spatially fixed mesh according to motion of the material particles. This requires the numerical solution of a hyperbolic type of partial differential equation, the so called advection equation. An advantage of the fractional-step approach against unsplit methods is the easy implementation into an existing Lagrange code. In addition, each type of partial differential equations can be treated independently by specific designed numerical algorithms. While for the solution of the Lagrange step stable and efficient solution techniques are established, the numerical treatment of advection dominated problems is still under discussion also in the field of Computational Fluid Dynamics. The following subsection is dedicated for a systematic comparison of alternative numerical techniques for the solution of advection equations.

4.2 Numerical Methods for Advection Dominated Problems

The hyperbolic advection equation describes the time and space dependent transport of a known quantity $\boldsymbol{\alpha}(x,t)$ moving with the velocity $\boldsymbol{w}$ in spatially fixed observer framework. The advection process of a scalar variable α in one dimension is expressed as

$$\frac{\partial \alpha}{\partial t} + w\frac{\partial \alpha}{\partial x} = 0. \qquad (31)$$

This linear advection process describes the transport of the quantity α along a distance $w\Delta t$ without changing its magnitude.

Numerical approaches known from computational fluid mechanics can be classified into

- finite difference methods (e.g. Euler, Lax–Wendroff, Crank–Nicolson),
- finite volume methods (e.g. Godunov, Van Leer, κ-scheme) or
- finite element methods (e.g. Galerkin methods, Discontinuous Galerkin methods (DG)).

For details the reader is referred to [1] and [12]. Many of the numerical schemes suffer from problem dependent specific numerical inaccuracies like diffusive or dispersive effects as well as oscillatory response. A discussion of the numerical accuracy of different schemes for the treatment of hyperbolic advection dominated problems is outlined below.

Finite difference methods (FDM) are the most traditional approaches for the numerical treatment of advection equations. Here only the most common schemes will be analyzed, a survey on the variety of these schemes is given in [21]. For FDM it is distinguished between implicit methods, e.g. backward Euler $\mathcal{O}(\Delta t)$, Crank–Nicolson $\mathcal{O}(\Delta t^2)$, and explicit methods, e. g. forward Euler $\mathcal{O}(\Delta t)$, forward Euler upstream $\mathcal{O}(\Delta t)$, Lax–Friedrich $\mathcal{O}(\Delta t)$, Lax–Wendroff $\mathcal{O}(\Delta t^2)$ or Leap-Frog $\mathcal{O}(\Delta t^2)$.

The explicit methods must satisfy the Courant condition

$$Cr = \frac{w\Delta t}{\Delta x} \leq 1 \qquad (32)$$

for conditional stability. Illustratively, the Courant criterion ensures that particles do not skip elements on their way through the mesh. The use of FDM approaches for three dimensional rolling contact problems within the ALE framework is limited by this stability condition, especially when unstructured spatial discretization is necessary.

For an overview on finite volume methods (FVM) for the treatment of advection equations, see [34]. Basis for this approach is the integral-form of (31), where the gradient term is transformed into a boundary integral,

$$\int_\Omega \frac{\partial \alpha}{\partial t}\, \mathrm{d}V + \int_{\Gamma_e} \alpha\, \boldsymbol{n} \cdot \boldsymbol{w}\, \mathrm{d}\Gamma = \int_\Omega \frac{\partial \alpha}{\partial t}\, \mathrm{d}V + F_e(\alpha, \boldsymbol{w}, \boldsymbol{n})$$
$$= \int_\Omega \frac{\partial \alpha}{\partial t}\, \mathrm{d}V + \sum_{k=1}^{n} h^k(\alpha, \boldsymbol{w}^k) = 0. \qquad (33)$$

The boundary term describes the weighted flux F_e of the internal variables streaming over the n boundaries of the domain, e.g. a finite volume cell. The accuracy and numerical stability depends on the choice of the functions h^k as approximation of the partial fluxes. For applications in fluid mechanics different techniques have been introduced, e.g. Godunov (total upwinding) [14], κ-scheme with Minmod- and Superbee-Limiters [34], Van Leer [34] or Lax–Wendroff [28]. The assumption that all relevant information comes from the upstream direction is called "upwinding". In classical upwind schemes only the direct neighbor elements are taken into account. The application of a total upwind scheme for the computation of a rolling viscoelastic wheel has been reported in [23].

Time-Discontinuous-Galerkin (TDG) methods have been introduced more recently for the solution of advection dominated problems, see [2, 8] for example. By this approach a space-time-finite element discretization is used instead the common concept semi-discretization. While standard (C^0-smooth) shape functions are used for the discretization in space, discontinuous (C^{-1}-smooth)

Finite Element Techniques for Rolling Rubber Wheels 137

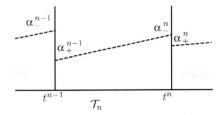

Fig. 5 Linear temporal discontinuous finite element approximation.

shape functions are used for temporal discretization within a time interval $\mathcal{T}_n = [t^{n-1}, t^n]$.

The TDG scheme is based on a weak formulation of the advection equation and yields to A-stability for linear PDEs, cf. [19]. The local form is multiplied by a test function η and integrated both in time and space,

$$\int_\Omega \int_{\mathcal{T}_n} \left(\eta \frac{\partial \alpha}{\partial t} + \eta w \frac{\partial \alpha}{\partial x} \right) \mathrm{d}t \, \mathrm{d}V = 0. \qquad (34)$$

Because of the jumps $[|\alpha|] = \alpha_+^{n-1} - \alpha_-^{n-1}$ at the time interval boundaries two different values of the function $\alpha(x,t)$ are computed at time t_n, see Figure 5,

$$\alpha_+^n = \lim_{s \to 0} \alpha(t^n + s), \quad \alpha_-^n = \lim_{s \to 0} \alpha(t^n - s). \qquad (35)$$

Due to those discontinuities at the interval boundaries initial conditions are necessary for each new time step. Choosing the results α_-^{n-1} of the previous time step $\mathcal{T}_{n-1}$ as initial values for the current interval leads to an implicit integration scheme, resulting into the rewritten weak formulation

$$\int_\Omega \int_{\mathcal{T}_n} \left(\eta \frac{\partial \alpha}{\partial t} + \eta w \frac{\partial \alpha}{\partial x} \right) \mathrm{d}t \, \mathrm{d}V + \int_\Omega \eta_+^{n-1} \alpha_+^{n-1} \, \mathrm{d}V = \int_\Omega \eta_+^{n-1} \alpha_-^{n-1} \, \mathrm{d}V. \qquad (36)$$

The function $\alpha(x,t)$ and the test function $\eta(x,t)$ are approximated by Lagrangian polynomials in time $N(t)$ and in space $H(x)$,

$$\begin{aligned} \alpha(x,t) &= N(t) \, \tilde{\alpha}(x) = N(t) \, H(x) \, \hat{\alpha}, \\ \eta(x,t) &= N(t) \, \tilde{\eta}(x) = N(t) \, H(x) \, \hat{\eta}. \end{aligned} \qquad (37)$$

The order of the integration scheme depends on the order of the polynomials for temporal approximation within the time interval,

$$\begin{aligned} \text{TDG}(0) &\to N(t) : \text{constant} &\Rightarrow \mathcal{O}(\Delta t), \\ \text{TDG}(1) &\to N(t) : \text{linear} &\Rightarrow \mathcal{O}(\Delta t^3), \\ \text{TDG}(2) &\to N(t) : \text{quadratic} &\Rightarrow \mathcal{O}(\Delta t^5). \end{aligned}$$

Thus, TDG schemes provide opportunities to enhance the accuracy by simply choosing higher order polynomials. However, with increasing polynomial order the computational effort increases mentionable. With the approximation (37) the discretized form of (34) is obtained,

$$\underbrace{\left[\int_{\mathcal{I}_n} N^T N_{,t}\, dt + N_+^T(t^{n-1})N_+(t^{n-1})\right]}_{\Upsilon_a} \int_{\Omega_e} \tilde{\eta}\,\tilde{\alpha}\, dV \tag{38}$$

$$+ \underbrace{\left[\int_{\mathcal{I}_n} N^T N\, dt\right]}_{\Upsilon_b} \int_{\Omega_e} \tilde{\eta}\, w\, \nabla \tilde{\alpha}\, dV = \underbrace{\left[N_+^T(t^{n-1})N_-(t^{n-1})\right]}_{\Upsilon_c} \int_{\Omega_e} \tilde{\eta}\,\tilde{\alpha}^{n-1}\, dV.$$

The subsequent spatial discretization based on a standard (continuous) Bubnov–Galerkin approach for $\tilde{\eta}$ and $\tilde{\alpha}$ leads to a non-symmetric linear system coupled in time and space,

$$[\Upsilon_a(t)\,\mathbf{M}(x) + \Upsilon_b(t)\,\mathbf{Q}(x)]\,[\hat{\alpha}] = [\Upsilon_c(t)\,\mathbf{M}(x)\,\hat{\alpha}^{n-1}]. \tag{39}$$

Herein, Υ_i describe the temporal approximation matrices, $\mathbf{M}$ has the structure of a standard mass matrix, and $\mathbf{Q}$ is the non-symmetric advection matrix. In detail, these matrices are written as

$$\mathbf{M} = \int_{\Omega_e} \mathbf{H}^T \mathbf{H}\, dV \quad \text{and} \quad \mathbf{Q} = \int_{\Omega_e} w\, \mathbf{H}^T \mathbf{H}_{,\chi}\, dV, \tag{40}$$

where $\mathbf{H}$ is the matrix of standard finite element spatial shape functions, and $\mathbf{H}_{,\chi}$ represents the gradient with respect to the reference configuration.

For the evaluation of the time matrices $\Upsilon_i(t)$ the global time interval $\mathcal{I}_n = [t_{n-1}, t_n]$ is projected onto the local interval $\mathcal{I}_n = [-1, 1]$ with the local time coordinate $\vartheta(t)$. This step is comparable to the mapping within an isoparametric concept. By the transformation

$$\vartheta(t) = a_1 + a_2\, t \quad \text{with} \tag{41}$$
$$\vartheta(t^{n-1}) = -1 = a_1 + a_2\, t^{n-1}, \quad \vartheta(t^n) = 1 = a_1 + a_2\, t^n$$

the local time coordinate $\vartheta(t)$ follows to

$$\vartheta(t) = \frac{1}{\Delta t}\left(2t - t^{n-1} - t^n\right) \quad \Rightarrow \quad \vartheta_{,t} = \frac{2}{\Delta t}, \tag{42}$$

and the inverse transformation to

$$t = \frac{1}{2}\left(\Delta t\, \vartheta + t^{n-1} + t^n\right) \quad \text{with}$$
$$t_{,\vartheta} = \frac{\Delta t}{2} \quad \Rightarrow \quad dt = t_{,\vartheta}\, d\vartheta = \frac{\Delta t}{2}\, d\vartheta. \tag{43}$$

Using (42) and (43) the time matrices can be computed directly by evaluating the time integral for the unit interval $\mathcal{I}_n$,

$$\Upsilon_a = \int_{\mathcal{I}_n} N^T(t)\, N_{,t}(t)\, dt + N_+^T(t^{n-1}) N_+(t^{n-1})$$

$$= \int_{-1}^{1} N^T(\vartheta)\, N_{,\vartheta}(\vartheta)\, d\vartheta + N^T(\vartheta=-1) N(\vartheta=-1), \quad (44)$$

$$\Upsilon_b = \int_{\mathcal{I}_n} N(t)^T\, N(t)\, dt = \frac{\Delta t}{2} \int_{-1}^{1} N^T(\vartheta) N(\vartheta)\, d\vartheta, \quad (45)$$

$$\Upsilon_c = N_+^T(t^{n-1}) N_-(t^{n-1}) = N^T(\vartheta=-1) N(\vartheta=1). \quad (46)$$

The size of the linear system (39) strongly depends on the polynomial order chosen for temporal approximation. For a constant approximation within the time-interval (TDG(0) scheme) for example, the matrices Υ_i degrade to scalar values and the dimension of the the linear system is $n \times n$, where n describes the number of nodes of the spatial discretization. A linear approximation within the time interval (TDG(1) scheme) leads to a system of size $2n \times 2n$, and so on. The temporal matrices for these cases are given below:

$$\text{TDG}(0): \quad \Upsilon_a = 1, \quad \Upsilon_b = \Delta t, \quad \Upsilon_c = 1, \quad (47)$$

$$\text{TDG}(1): \quad \Upsilon_a = \frac{1}{2}\begin{bmatrix} 1 & 1 \\ -1 & 1 \end{bmatrix}, \quad (48)$$

$$\Upsilon_b = \frac{\Delta t}{6}\begin{bmatrix} 2 & 1 \\ 1 & 2 \end{bmatrix}, \quad \Upsilon_c = \begin{bmatrix} 0 & 1 \\ 0 & 0 \end{bmatrix},$$

$$\text{TDG}(2): \quad \Upsilon_a = \frac{1}{6}\begin{bmatrix} 3 & 4 & -1 \\ -4 & 0 & 4 \\ 1 & -4 & 3 \end{bmatrix}, \quad (49)$$

$$\Upsilon_b = \frac{\Delta t}{30}\begin{bmatrix} 4 & 2 & -1 \\ 2 & 16 & 2 \\ -1 & 2 & 4 \end{bmatrix}, \quad \Upsilon_c = \begin{bmatrix} 0 & 0 & 1 \\ 0 & 0 & 0 \\ 0 & 0 & 0 \end{bmatrix}.$$

4.3 Comparison of Numerical Advection Schemes

In this subsection the behavior of the different approaches for the numerical treatment of advection problems are presented. In a first example the numerical results are compared with the analytical solution of the 1D-advection problem for a Gaussian profile with initial condition

$$\alpha(x, t=0) = \exp\left[-\frac{(x+2)^2}{2\sigma}\right]$$

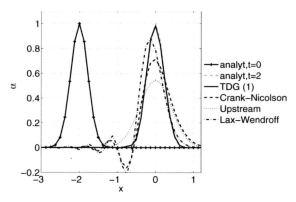

Fig. 6 Advection of Gauss profile: Comparison of FD and TDG schemes.

and the analytical solution

$$\alpha(x, t > 0) = \exp\left[-\frac{(x + 2 - w\,t)^2}{2\sigma}\right].$$

The model parameters are described by a standard deviation of $\sigma = 0.05$ and a convective velocity of $w = 1$ m/s. These parameters yield into a very steep but continuous function with large gradients. Some results are depicted in Figure 6 to 8 for different schemes and for different Courant numbers. In Figure 6 concurrent FD schemes are compared with the linear TDG approach for a subcritical Courant number $Cr = 0.53$. Within the FD family, upstream and Lax–Wendroff schemes seem to be the most accurate approaches. The upstream approach is stable, but suffers strongly from diffusion effects, a much better conservation of the amplitude is observed from Lax–Wendroff and Leap-Frog methods. However, the latter tend to oscillations. The most accurate results obtained with FD methods have been for Courant numbers near to one, from which the requirement of an adaptive time stepping scheme with respect to the spatial discretization is concluded. In contrast, the TDG(1) approach shows neither oscillatory nor dispersive behavior.

In contrast to the FD methods the TDG methods are not limited by the Courant condition (32) and present even for larger time steps ($Cr > 1$) stable solutions. The results for TDG schemes of different polynomial order are compared in Figure 7 for a Courant number $Cr = 2.66$, where the behavior of improved approximation with increasing polynomial order is evidently observed.

As an example for the transport of a discontinuous function a square pulse with infinite gradients is analyzed. With the initial conditions

$$\alpha(x, t = 0) = \begin{cases} 1 & : \quad -1 \leq x \leq 1, \\ 0 & : \quad \text{else,} \end{cases}$$

Finite Element Techniques for Rolling Rubber Wheels 141

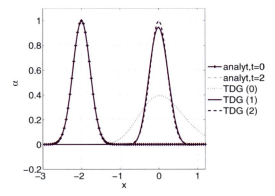

Fig. 7 Advection of Gauss profile: Comparison of p-order for TDG schemes.

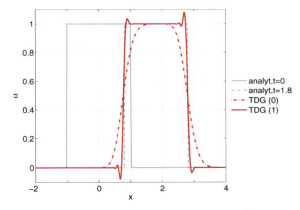

Fig. 8 Advection of a discontinuous function ($Cr = 1.44$).

this problem is described by the analytical solution

$$\alpha(x, t > 0) = \begin{cases} 1 & : \quad -1 + wt \leq x \leq 1 + wt, \\ 0 & : \quad \text{else.} \end{cases}$$

The computational results depicted in Figure 8 show that, in principle, it is possible to compute such steep fronts by using the presented algorithm, but oscillations appear in the regions of the discontinuities. These oscillations are known as Gibbs' phenomenon and they are explained by so called over and undershoots. These effects result from differences between the inflow and the outflow in one element. They can be avoided when choosing the TDG(0) scheme, e.g. constant temporal approximation, but this scheme shows diffusive behavior. For the avoidance or at least the reduction of over and undershoots, stabilization techniques have been investigated [30]. For this kind of problems in fluid dynamics alternatively spatially DG methods have been suggested, see [8], which allow for jumps at the element boundaries.

4.4 Numerical Benchmark

For the treatment of inelastic material behavior the proposed TDG approach for the solution of the advection equation is incorporated into an algorithm for rolling contact analysis within the ALE framework. By the use of the fractional-step approach an explicit scheme is introduced. In each time step the following steps have to be performed:

1. *Lagrange Step*: Solution of the nonlinear ALE rolling contact problem (5) at time $t = t_a$ by neglecting convective terms.
2. *Projection Step*: Smoothing of internal variables for a C^0-smooth representation.
3. *Euler Step*: Transport of internal variables by solving the advection equation by TDG schemes (39).

In the first step the mechanical balance equation for the nonlinear rolling contact problem is computed by solving the linearized finite element approximation (5) within a Newton–Raphson scheme. After the mechanical equilibrium is obtained, the evolution rules for the internal variables are integrated locally, which is referred as Lagrangian step within the fractional step approach.

For the advection step a continuous field of the internal variables has to be provided, whose values have been computed at distinct integration points inside the finite elements so far. Well established *super-convergent patch recovery* techniques are applied for the projection of the history variables onto the nodal points of the spatial finite element mesh to obtain a C^0-smooth representation.

In the final Eulerian step, the advection equation (39) is solved for the transport of the material history within the spatially fixed mesh. The mechanical equilibrium is reiterated simultaneously.

The overall algorithm is based on a consistent mathematical description and there is no need for structured spatial finite element meshes. It is applicable for arbitrary inelastic constitutive equations based on thermodynamical consistent internal variable description.

As numerical example a finite element model for a Grosch wheel has been studied. The discretized wheel is depicted in Figure 9 in a deformed contact status rolling on a rigid flat surface. Its initial radius is about 40 mm and the thickness is 10 mm. The spatial discretization consists of about 3400 eight-node non-linear brick elements and in addition of about 300 contact elements resulting about 10,000 unknowns. The inner ring is described with Dirichlet conditions ($u_x = 4$ mm, $u_y = u_z = 0$). The constitutive description has been given according to Section 3, while the so called $\bar{B}$-concept has been chosen

Finite Element Techniques for Rolling Rubber Wheels

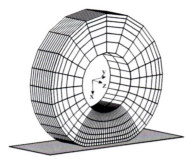

Fig. 9 Finite element model of a Grosch wheel.

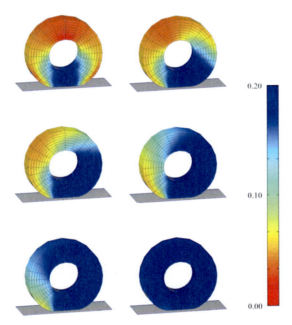

Fig. 10 Transport of an scalar valued dimensionless isotropic damage ($0 \leq D \leq 1$) variable through the spatially fixed FE-mesh.

for tackling the volumetric contributions. The contact itself is assumed to be frictionless in order not to mix up additional dissipative effects.

A first result is shown in Figure 10 in which the temporal evolution of a scalar valued internal variable for isotropic damage is depicted for six time steps. The local damage evolution obviously is initiated in the contact domain and the damaged constitutive properties are transported according to the material motion in the spatially fixed finite element mesh. Thus the damage spreads circumferentially until all material point at same concentric positions are damaged equally in the stationary rolling stage.

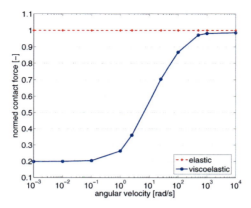

Fig. 11 Resulting contact force of a viscoelastic Grosch wheel depending on rolling speed.

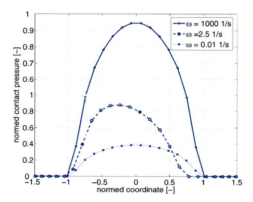

Fig. 12 Contact pressure distribution of a traction-less rolling viscoelastic Grosch wheel for different rolling speeds.

In a second example explicitly time dependent viscoelastic effects have been studied. In the context of the non-linear constitutive theory outlined in Section 3, a viscoelastic model with only one relaxation time has been considered here for purpose of verification. The wheel is driven with controlled displacement against the rigid plane road while the rolling speed is increased.

In Figure 12 the computed contact force is drawn in dependence of the rolling speed. For small velocities only a moderate contact force is obtained which rapidly increases with increasing rolling speed and converges to a constant value again. This is in clear correspondence with the constitutive model to be interpreted in analogy to a three parameter linear viscoelastic rheological model. At quite low speed the dashpot is inactive and therefore, only a parallel spring is active. At higher speed the dashpot gets active and the

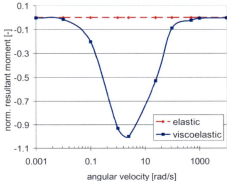

Fig. 13 Rolling resistance moment computed for a Grosch wheel depending on speed.

stiffness increases while for very high speed the dashpot behaves rigid and the full stiffness of two parallel springs is present.

The corresponding contact pressure distribution for a concentric ring of contact nodes is depicted in Figure 11 for three different speeds. While at quite low and rather high speed a symmetric shaped contact pressure distribution is obtained, the pressure maximum tends to the leading edge of the footprint resulting into a rolling resistance torque shown in Figure 13.

5 Treatment of Friction within the ALE Formulation of Rolling Bodies

The kinematical model for a rolling wheel is sketched in Figure 14. The wheel is spinning with angular velocity ω around a spatially fixed axis while the road is moving with constant ground velocity v_F. Due to the axis load the initial radius r_0 is flattened to r. For any material particle of the wheel the guiding velocity $\boldsymbol{w}$ and the convective velocity $\boldsymbol{c}$ can be easily computed from (2).

The normal contact constraints are treated as spatial quantities. Related formulations and computational methods, cf. [22, 36], are directly adapted from Lagrangian mechanics, we will not repeat them here. But the tangential contact constraints have to be formulated in a different manner, because the history of two contacting particles moving through the finite contact area are not computed directly within the spatially fixed ALE mesh.

The stick conditions in the material picture are described by $s_\alpha = 0$, where s_α, $\alpha = 1, 2$, describes the spatial coordinates of the relative tangential displacement of two contacting particles, referred to as local slip. The temporal derivative $\dot{s}_\alpha$ is called local slip velocity. The stick condition in general reads,

$$s_\alpha = 0 \quad \text{and} \quad \dot{s}_\alpha = 0. \tag{50}$$

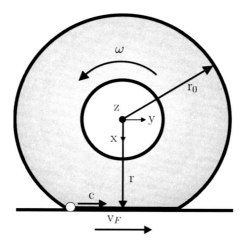

Fig. 14 Kinematics of rolling.

While the slip velocity is computed easily within the ALE description of rolling wheels, the determination of the slip itself requires particle tracking methods similar to the treatment of history dependent material behavior. Therefore, the path-lines of the contacting particles have to be reconstructed from the known velocity field $c_\alpha = \boldsymbol{c} \cdot \bar{\boldsymbol{a}}_\alpha$, where $\bar{\boldsymbol{a}}_\alpha$ denote the unit vectors tangential to the contact surface. In a componentwise notation the integration along the path coordinate ζ then follows as

$$s_\alpha(\zeta) = s_{0\alpha} + \int \dot{s}_\alpha \, dt = s_{0\alpha} + \int_{\zeta_0}^{\zeta} \frac{\dot{s}_\alpha(\bar{\zeta})}{c_\alpha(\bar{\zeta})} \, d\bar{\zeta} \qquad (51)$$

with $d\bar{\zeta} = c_\alpha(\zeta) \, dt$.

In the special case where a deformable wheel is rolling on a rigid road the slip velocities are computed as the difference between the tangential projection of the convective velocity $\boldsymbol{c}$ and the prescribed ground velocity $\boldsymbol{v}_F$,

$$\dot{s}_\alpha = (\boldsymbol{c} - \boldsymbol{v}_F) \cdot \bar{\boldsymbol{a}}_\alpha = (\operatorname{Grad} \boldsymbol{\varphi} \cdot \boldsymbol{w} - \boldsymbol{v}_F) \cdot \bar{\boldsymbol{a}}_\alpha. \qquad (52)$$

The convective velocities, defined by (2), depend on the displacement gradient. They are evaluated at the points of superconvergence within a C^0-smooth approximation. Under sticking conditions the velocities of material points of the wheel are equal to the ground velocity.

For the efficient numerical evaluation of the described problem, (51) is not solved directly. Recalling the material time derivative (2) the slip velocity in steady-state motion can be expressed as

$$\dot{s}_\alpha = \operatorname{Grad} s_\alpha \cdot \boldsymbol{w} \,. \qquad (53)$$

This proposed ansatz requires a C^0-smooth distribution of the nodal slip variables s_α for which additional variables in the active part of the contact area are introduced. The combination of (52) and (53) leads to

$$\text{Grad}\, s_\alpha \cdot \boldsymbol{w} = (\text{Grad}\, \boldsymbol{\varphi} \cdot \boldsymbol{w} - \boldsymbol{v}_F) \cdot \bar{\boldsymbol{a}}_\alpha. \tag{54}$$

For a proper physical description of the rolling conditions the entry condition of the material points has to be added, cf. (51). For stationary rolling the assumption is stated that the particles enter the contact zone smoothly, without any relative tangential displacement. Thus, homogeneous Dirichlet conditions are applied to the nodes at the leading edge of the contact area,

$$s_\alpha = 0 \quad \text{at} \quad \Gamma_0 \subset \partial_c \Phi(B), \tag{55}$$

where $\partial_c \Phi(B)$ denotes the contact area in the current configuration. At the trailing edge the particles leave the contact zone without any restraints. In this part of the contact zone the so called *snap-out* effect often is observed, where the particles are accelerated back to their initial path velocity.

For the computation of the slip variables s_α we start from the weak representation of (54),

$$\int_{\partial_c \Phi(B)} (\text{Grad}\, s_\alpha \cdot \boldsymbol{w})\, \delta \dot{s}^\alpha\, \mathrm{d}a = \int_{\partial_c \Phi(B)} (\text{Grad}\, \boldsymbol{\varphi} \cdot \boldsymbol{w} - \boldsymbol{v}_F) \cdot \bar{\boldsymbol{a}}_\alpha\, \delta \dot{s}^\alpha\, \mathrm{d}a. \tag{56}$$

Herein, $\delta \dot{s}_\alpha$ represents the material variation of the slip velocity $\dot{s}_\alpha$ in the tangential plane. For the solution of the nonlinear problem within an incremental iterative scheme (56) has to be linearized with respect to both, the nodal slip variables s_α and the displacement variables $\boldsymbol{\varphi}$,

$$\int_{\partial_c \Phi} [\text{Grad}\, \Delta s_\alpha \cdot \boldsymbol{w} - (\text{Grad}\, \Delta \boldsymbol{\varphi} \cdot \boldsymbol{w}) \cdot \bar{\boldsymbol{a}}_\alpha]\, \delta \dot{s}^\alpha\, \mathrm{d}a$$
$$= \int_{\partial_c \Phi} [(\text{Grad}\, {}^t\boldsymbol{\varphi} \cdot \boldsymbol{w} - \boldsymbol{v}_F) \cdot \bar{\boldsymbol{a}}_\alpha - \text{Grad}\, {}^t s_\alpha \cdot \boldsymbol{w}]\, \delta \dot{s}^\alpha\, \mathrm{d}a. \tag{57}$$

From (57) the slip distribution in the contact area is computed directly. Once the spatial slip distribution is computed, standard concepts for the treatment of frictional contact problems can be applied immediately, see e.g. [35].

For the computation of the tangential shear traction $\boldsymbol{\tau}$ it has to be distinguished between stick and slip. Assuming Coulomb's law for simplicity, the friction function reads

$$f = \|\boldsymbol{\tau}\| - \tau_{\max} \leq 0 \quad \text{with} \quad \tau_{\max} = \mu\, p, \tag{58}$$

where μ is the constant friction coefficient. The stick-slip distinction is expressed by

$$f < 0 \quad \text{and} \quad s_\alpha = 0 \quad \text{(stick)},$$
$$f = 0 \quad \text{and} \quad s_\alpha > 0 \quad \text{(slip)}. \tag{59}$$

In this strong mathematical setting no relative displacement s_α between the contacting particles is allowed in the sticking case. In a regularized form of the friction law small reversible slip is tolerated, which is physically motivated by elastic micro deformation of surface asperities. Thus, the total slip distance splits into a small elastic (sticking) part and an irreversible (sliding) contribution,

$$s_\alpha = s_\alpha^{el} + s_\alpha^{irr} \quad \text{and} \quad \dot{s}_\alpha = \dot{s}_\alpha^{el} + \dot{s}_\alpha^{irr}. \tag{60}$$

For the elastic part s_α^{el} a linear constitutive dependency

$$s_\alpha^{el} = -\frac{1}{\epsilon_t} \tau_\alpha \quad \Leftrightarrow \quad \tau_\alpha = -\epsilon_t\, s_\alpha^{el} \tag{61}$$

is assumed, which can be interpreted as a penalty regularization of the stick conditions.

In the case of sliding, when the friction law is violated and the initial assumption of sticking has to be neglected, the irreversible part s_α^{irr} of the total slip is computed from the evaluation of the non-associative friction rule

$$\dot{s}_\alpha^{irr} = \dot{\lambda}\, \frac{\partial f}{\partial \tau_\alpha} = \dot{\lambda}\, \frac{\tau_\alpha}{\|\tau\|} = \dot{\lambda}\, n_{t\alpha}, \tag{62}$$

which can be integrated directly by the use of radial return-mapping schemes.

The frictional rolling contact conditions are included into the weak form of the equation of motion (3) by adding the term of the virtual work of the tangential contact forces,

$$C = -\int_{\partial_c \Phi(B)} \tau_\alpha\, \delta s^\alpha\, da \quad \text{with} \quad \delta s_\alpha = \delta\boldsymbol{\varphi} \cdot \bar{\boldsymbol{a}}_\alpha. \tag{63}$$

In contrast to the material variation of the slip velocity in (56), δs_α is a spatial variation of the current location of the particle.

The linearized form of (63) reads

$$\int_{\partial_c \Phi(B)} {}^{t+\Delta t}\tau_\alpha\, \delta s^\alpha\, da = \int_{\partial_c \Phi(B)} {}^{t}\tau_\alpha\, \delta s^\alpha\, da + \int_{\partial_c \Phi(B)} {}^{t}\tau_\alpha\, \Delta \delta s^\alpha\, da \\ + \int_{\partial_c \Phi(B)} \Delta\tau_\alpha\, \delta s^\alpha\, da + \int_{\partial_c \Phi(B)} {}^{t}\tau_\alpha\, \delta s^\alpha\, \Delta da, \tag{64}$$

where $\Delta \delta s_\alpha = 0$ for the contact with a rigid flat surface because related curvature terms vanish.

The non-smoothness in the tangential contact law is treated by a predictor-corrector scheme. In the predictor step a trial state is computed under the assumption of sticking

$$^{t+\Delta t}\tau_\alpha^{\text{trial}} = {}^{t}\tau_\alpha - \epsilon_t \Delta s_\alpha = -\epsilon_t \left({}^{t+\Delta t}s_\alpha - {}^{t}s_\alpha^{\text{irr}}\right). \tag{65}$$

If the stick condition $(f(\boldsymbol{\tau}^{\text{trial}}) < 0)$ is violated a corrector step is carried out

$$\begin{aligned}{}^{t+\Delta t}\tau_\alpha &= {}^{t+\Delta t}\tau_\alpha^{\text{trial}} - \epsilon_t \Delta\lambda \, {}^{t+\Delta t}n_{t\,\alpha} \\ &= {}^{t+\Delta t}\tau_\alpha^{\text{trial}} - f(\boldsymbol{\tau}^{\text{trial}}) \, {}^{t+\Delta t}n_{t\,\alpha}.\end{aligned} \tag{66}$$

The incremental tangential contact stress update yields to

$$\Delta\boldsymbol{\tau} = -\epsilon_t \Delta \boldsymbol{s} \tag{67}$$

in the case of local stick, and

$$\Delta\boldsymbol{\tau} = \epsilon_n \mu \, \Delta d_n \, \boldsymbol{n}_t + \underbrace{\epsilon_t \frac{\mu p}{\|\boldsymbol{\tau}^{\text{trial}}\|} \left(\mathbf{1} - \boldsymbol{n}_t \otimes \boldsymbol{n}_t\right)}_{\epsilon_t^{\text{fric}}} \Delta \boldsymbol{s} \tag{68}$$

in the case of slip. Herein Δd_n describes the incremental rate of the normal penetration.

In our implementation the contact area at the surface of the rolling body is discretized with four-node contact-elements. Because of the restriction to the deformable-rigid contact only the *slave*-body is discretized. Within a *segment-to-surface* concept the contact stresses are computed at the integration points. In the contact patch the global displacement field $\boldsymbol{\varphi}$ and the spatial slip variables s_α are approximated by bilinear shape functions $H^i(\xi,\eta) = \frac{1}{4}\left(1 + \xi\xi_i\right)\left(1 + \eta\eta_i\right)$,

$$\begin{aligned}\boldsymbol{\varphi}(\xi,\eta) &= \sum_{i=1}^{4} H^i(\xi,\eta)\,\hat{\boldsymbol{\varphi}}_i = \mathbf{H}\,\hat{\boldsymbol{\varphi}} \quad \text{and} \\ \boldsymbol{s}(\xi,\eta) &= \sum_{i=1}^{4} H^i(\xi,\eta)\,\hat{\boldsymbol{s}}_i = \mathbf{H}_s\hat{\boldsymbol{s}}\,,\end{aligned} \tag{69}$$

with the local coordinates ξ_i, η_i of the isoparametric reference element. In addition to the three nodal degrees of freedom for the displacements there are two nodal degrees of freedom for the slip variables s_α, to be switched with the active set of contact. The related element matrices are labeled by the indices φ and s.

The finite element discretization of the global displacement field $\boldsymbol{\varphi}$ and the spatial slip variables s_α yields a coupled system for the linearized ALE formulation of the rolling contact problem,

$$\begin{bmatrix} {}^{t}\mathbf{K} - \mathbf{W} + \mathbf{K}_{c_{\varphi\varphi}} & \mathbf{K}_{c_{\varphi s}} \\ \hline \mathbf{K}_{c_{s\varphi}} & \mathbf{K}_{c_{ss}} \end{bmatrix} \begin{bmatrix} \Delta\hat{\boldsymbol{\varphi}} \\ \hline \Delta\hat{\boldsymbol{s}} \end{bmatrix} = \begin{bmatrix} {}^{t+\Delta t}\mathbf{f}_{ext} + {}^{t}\mathbf{f}_i - {}^{t}\mathbf{f}_\sigma - {}^{t}\mathbf{f}_{c_\varphi} \\ \hline -{}^{t}\mathbf{f}_{c_s} \end{bmatrix}. \tag{70}$$

By this approach the solution of the frictional contact problem is coupled in the displacement field φ and the slip variable s. The four–node contact-element has twelve degrees of freedom for the displacements and additionally eight degrees of freedom for the nodal slip. In the following the matrices corresponding to the weak form of the tangential contact forces (63) and the computation of the nodal slip (56) will be derived. These matrices are labeled by the index c in (70). The subindices φ and s are used for the corresponding equations and degrees of freedom.

At first, Equation (57) for the computation of the total nodal slip is discretized,

$$\delta\hat{\mathbf{s}}^T \left({}^t\mathbf{K}_{c_{ss}} \Delta\hat{\mathbf{s}} + {}^t\mathbf{K}_{c_{s\varphi}} \Delta\hat{\varphi} + {}^t\mathbf{f}_{c_s} \right) = \quad (71)$$

$$\delta\hat{\mathbf{s}}^T \left(\int_{\partial_c \Phi(B)} \mathbf{A}_s^T \mathbf{A}_s \, da \, \Delta\hat{\mathbf{s}} - \int_{\partial_c \Phi(B)} \mathbf{A}_s^T \mathbf{a}_t^T \mathbf{A} \, da \, \Delta\hat{\varphi} \right.$$

$$\left. + \int_{\partial_c \Phi(B)} \mathbf{A}_s^T \left[({}^t\mathbf{c} - \mathbf{v}_F) \, \mathbf{a}_t - \mathbf{A}_s \, {}^t\hat{\mathbf{s}} \right] da \right).$$

In this equation the following matrix representations are used for the description of the velocities and the material variation $\delta\dot{\mathbf{s}}$ of the sliding velocity,

$$\begin{aligned} \dot{\mathbf{s}} &= \text{Grad}\, s \cdot \boldsymbol{w} = \mathbf{H}_{s,\chi} \, \mathbf{w} \, \hat{\mathbf{s}} = \mathbf{A}_s \, \hat{\mathbf{s}}, \\ \delta\dot{\mathbf{s}} &= \mathbf{A}_s \, \delta\hat{\mathbf{s}} \quad \text{and} \quad \mathbf{c} = \text{Grad}\,\varphi \cdot \boldsymbol{w} = \mathbf{A}\,\hat{\varphi}. \end{aligned} \quad (72)$$

The only difference between the matrices $\mathbf{A}_s$ and $\mathbf{A}$ is the mapping onto two, respectively three degrees of freedom per element node. The matrix $\mathbf{a}_t$ represents the two tangential vectors $\bar{\boldsymbol{a}}_1, \bar{\boldsymbol{a}}_2$. In the case of local stick the discretized form of the linearized virtual work of the tangential contact forces (64) reads

$$\delta\hat{\varphi}^T \left({}^t\mathbf{f}_{c_\varphi} + {}^t \mathbf{K}_{c_{\varphi s}} \Delta\hat{\mathbf{s}} + {}^t \mathbf{K}_{c_{\varphi\varphi}} \Delta\hat{\varphi} \right) \quad (73)$$

$$= \delta\hat{\varphi}^T \left(-\int_{\partial_c \Phi(B)} \mathbf{H}^T \mathbf{a}_t {}^t\boldsymbol{\tau} \, da + \int_{\partial_c \Phi(B)} \epsilon_t \mathbf{H}^T \mathbf{a}_t \mathbf{H}_s \, da \, \Delta\hat{\mathbf{s}} \right.$$

$$\left. - \int_{\partial_c \Phi(B)} \left(\mathbf{H}^T \mathbf{a}_t {}^t\boldsymbol{\tau} \mathbf{H}_{,\chi} \right) da \, \Delta\hat{\varphi}^e \right).$$

Herein the spatial variation of the sliding distance is given by

$$\delta\mathbf{s}^T = \delta\hat{\varphi}^T \mathbf{H}^T \mathbf{a}_t. \quad (74)$$

In the case of local slip the tangential stress depends on the chosen friction law. For Coulomb's law the specific algorithmic consistent tangent operator $\boldsymbol{\epsilon}_t^{\text{fric}}$ according to (68) is applied,

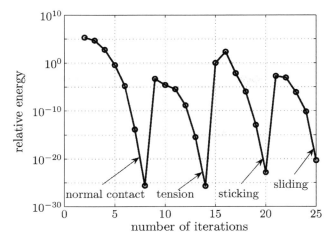

Fig. 15 Convergence behavior of a sequential rolling contact analysis.

$$\delta\hat{\varphi}^T \left({}^t\mathbf{f}_{c_\varphi} + {}^t\mathbf{K}^e_{c_{\varphi s}} \Delta\hat{\mathbf{s}} + {}^t\mathbf{K}^e_{c_{\varphi \varphi}} \Delta\hat{\varphi} \right) \tag{75}$$

$$= \delta\hat{\varphi}^T \left(-\int_{\partial_c \Phi(B)} \mathbf{H}^T \mathbf{a}_t {}^t\boldsymbol{\tau} \, \mathrm{d}a + \int_{\partial_c \Phi(B)} \mathbf{H}^T \mathbf{a}_t \boldsymbol{\epsilon}_t^{fric} \mathbf{H}_s \mathrm{d}a \Delta\hat{\mathbf{s}} \right.$$

$$\left. + \int_{\partial_c \Phi(B)} \left(\epsilon_n \mu \, \mathbf{n}_t \mathbf{H}^T \mathbf{a}_t \mathbf{a}_n^T \mathbf{H} - \mathbf{H}^T \mathbf{a}_t \, {}^t\boldsymbol{\tau} \mathbf{H}_{,\chi} \right) \mathrm{d}a \, \Delta\hat{\varphi}^e \right).$$

The assemblage of the element matrices leads to the coupled system of equations (70). This system is non-symmetric because $\mathbf{K}_{c_{\varphi s}} \neq \mathbf{K}_{c_{s\varphi}}^T$ and its size depends on the active set because of the additional nodal slip variables.

Concerning an efficient solution of the non-symmetric system of coupled equations (70) the possibility of a staggered solution neglecting the strong coupling has been evaluated, but only poor or none convergence of the iterative scheme has been obtained. In contrast, the fully coupled solution of the system yields quadratic convergence rates. The typical convergence behavior of the computation of a tractive rolling state is shown in Figure 15 which will be discussed below.

6 Numerical Examples

This section is aimed to demonstrate the physical reliability of the proposed approach as well as the industrial applicability of the presented algorithm. In a first example the frictional rolling contact of the Grosch wheel already introduced in Section 4 is studied in detail. As a second example the rolling contact behavior of a tire model is discussed.

6.1 Grosch Wheel

For testing the proposed algorithmic approach for frictional rolling, the Grosch wheel as described in Section 4.4 with the assumption of purely elastic constitutive properties, i.e. a Mooney–Rivlin model with the coefficients $c_{10} = 1$ MPa, $c_{01} = 0.67$ MPa and a bulk modulus $\kappa = 16.7$ MPa, has been investigated. A constant angular velocity $\omega = 10$ rad/s is assumed. The penalty-stiffness for the contact has been chosen to $\epsilon_n = 10^6$ MPa/mm for the normal and $\epsilon_t = 10^4$ MPa/mm in tangential direction. The friction coefficient for the Coulomb law has been chosen as $\mu = 0.5$.

In a first step the wheel is pressed displacement-controlled in one incremental step of $\Delta u = 4$ mm against the rigid plane contact surface. Friction is neglected in this step. Because of tensile contact stresses computed after equilibrium iteration the active set has been changed and the equilibrium has been re-iterated, this is seen as second arc in Figure 15. In a second step the tangential contact conditions are enforced by the approach introduced above. Again the active set is controlled conservatively, first complete sticking is assumed while friction is re-iterated when observed in the equilibrium state. The computation of both, the case of local stick and the case of local slip, yields to a quadratic rate of convergence as seen Figure 15.

The state of traction free rolling has been computed for a given ground velocity of $v_F = 397.2$ mm/s which has been found iteratively in the kinematical driven process. The obtained contact pressure distribution is shown in Figure 16a. In this example the homogeneity of typical parabolic shape is disturbed by high edge pressures which are the result of the restricted lateral motion of these particles due to finite friction. The corresponding tangential contact stresses are depicted in Figures 16b and c. In general the circumferential contact shear stress distribution appears to be S-shaped, as expected from analytical results. However, due to local effects observed at the lateral edges already discussed at the contact pressure distribution, the circumferential shear changes the sign at the edges. From the comparison of magnitudes between circumferential and lateral shear it is obvious that in this example the edge effects are dominated by the lateral contact shear traction depicted in Figure 16c.

Starting from the state of traction free rolling the ground velocity is increased to simulate braking. To quantify states of stationary braking a kinematic quantity called global slip is introduced

$$S = \frac{v_F - \omega r_o}{v_F}, \qquad (76)$$

which is measurable at the axis of the wheel. In Figure 17a the circumferential shear traction distribution and the motion within the footprint are depicted for stages of different braking slip. In the right column the total contact area is shaded, while the contact status (stick vs. slip) is indicated by different

Finite Element Techniques for Rolling Rubber Wheels

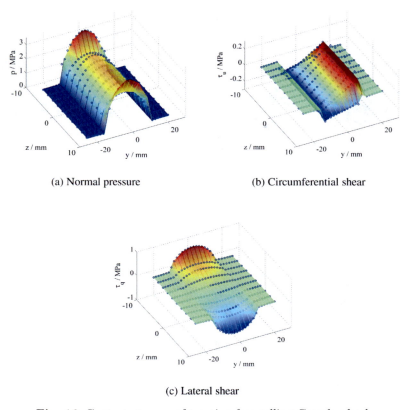

(a) Normal pressure

(b) Circumferential shear

(c) Lateral shear

Fig. 16 Contact stresses of traction free rolling Grosch wheel.

colors. In addition the local slip velocity as measure for the movement of the particles is shown by the small arrows in the center of each contact element.

In comparison to the free rolling state the shear traction increases at the trailing edge of the contact zone. With increasing global slip an increasing slip zone develops growing from the trailing to the leading edge. These results are in clear correspondence with analytical solutions [7]. Increasing the ground velocity leads to a rising tangential braking force going ahead with an increasing slip area until each element is in frictional contact.

The developed algorithms allow for the straight forward computation of more complicated contact conditions, like cornering situations as depicted in Figure 18. Here in the left the deformed mesh of the rolling Grosch wheel running under a cornering angle of $\alpha = 15°$ is shown. The computed contact pressure distribution is plotted in the right figure. Despite the contact algorithm for these computations works efficient and stable, model dependent instabilities have been observed by this example. The high edge contact pressure distribution found for the original model with rectangular cross section

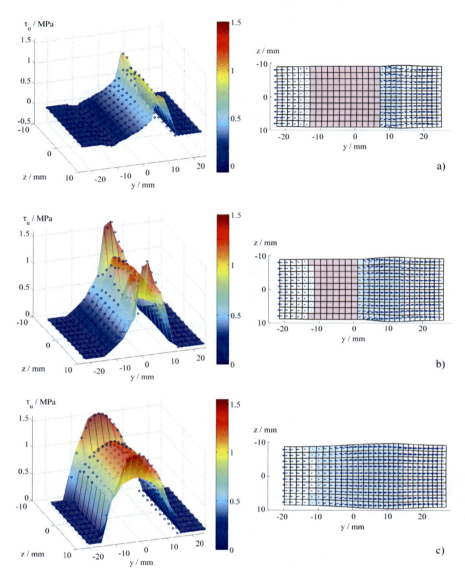

Fig. 17 Development of the circumferential shear traction and sliding zone for global breaking slip at (a) 4%, (b) 10% and (c) 30%.

leads locally to high shear traction in sliding case. At the boarder between sticking and sliding domain sharp kinks in the deformation pattern have been observed, leading to rather high distortions of the contact elements. This effect is reduced using rounded edges and refined spatial discretization of the contact domain.

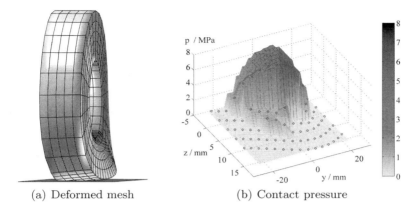

(a) Deformed mesh   (b) Contact pressure

Fig. 18 Grosch wheel under a cornering angle ($\alpha = 15°$).

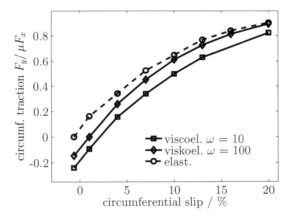

Fig. 19 Traction-slip diagram computed for the Grosch wheel with simple viscoelastic material and increasing braking slip.

For the analysis of combined dissipative effects the Grosch wheel with viscoelastic material behavior in combination with frictional rolling has been studied. The same viscoelastic model as discussed in Section 3 and increased circumferential slip have been applied. The traction-slip diagrams computed under these assumptions are depicted in Figure 19 for two different rolling speeds. The typically curved shape of the response results from the continuously increasing slip area, ending up into a horizontal line for total sliding. The vertical offset between elastic and viscoelastic analysis is explained by the speed dependent rolling resistance, observed already for the frictionless case as discussed in Section 4.4. Thus, viscoelastic bulk properties enhance

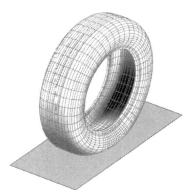

Fig. 20 FE model of a radial tire.

braking performance while the driving performance is decreased. Energy consumption in general is increased with increasing viscoelastic behavior.

6.2 Tire Model

To demonstrate the applicability of the developed approaches for industrial applications a detailed passenger car tire model with three circumferential grooves has been analyzed. The finite element model depicted in Figure 20 consists of about 45,000 degrees of freedom. The cross-section is discretized with 15 distinct material groups with different hyperelastic constitutive properties. The tire-model has been loaded with typical operational conditions, i.e. a vertical load of 4 kN, inflation with an internal pressure of 0.2 MPa and rotation corresponding to a rolling speed of 80 km/h. Coulomb's friction coefficient has been chosen to $\mu = 1.0$.

The contact stresses computed for the state of traction free rolling are depicted in Figure 21. In this state all contact contact elements stick. High pressure concentrations are observed at the lateral edges of the footprint which is typical for virgin radial tires as far as no wear and stress softening of the rubber material has taken place. In the circumferential direction the contact shear stress distribution appears S-shaped while no resultant force is transmitted. An interesting effect can be observed in lateral direction. Due to the grooves the lateral contact traction appears oscillating because the sticking particles are constrained for lateral spreading at the edges of each rib. This causes high local traction with changing directions in each rib. This behavior is in good qualitative correspondence with experimental results, cf. [6, 20].

When additional braking forces are applied to the tire a zone of sliding elements starts developing from the trailing edge to the leading edge. In Figure 22a the status of the contact elements and the slip velocities are plotted for a braking slip of $S = 5\%$. Again, sticking elements are shaded

Finite Element Techniques for Rolling Rubber Wheels

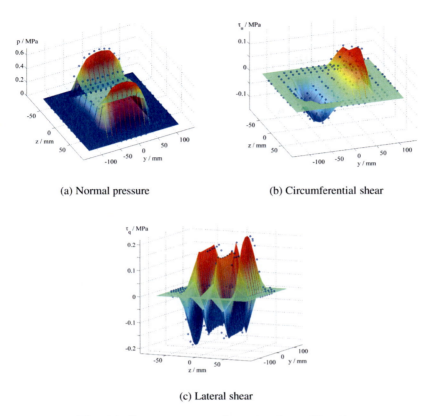

Fig. 21 Contact stresses of traction free rolling tire.

in dark grey and local sliding in bright grey. From Figure 22b it can be easily seen that the regions of maximal circumferential shear traction are in the regions of maximal normal pressure. Because of the assumed Coulomb friction, in the sliding areas the shape of the shear stress distribution clearly equals to that of the normal pressure.

The proposed algorithmic approach also enables for efficient analysis of cornering effects simply by adding a lateral component to the ground velocity v_F, corresponding to a slip angle δ. This has been studied using a slick-tire, illustrated by the computational results shown in Figure 23. With increasing slip angle the tread of the tire moves to the outside of the curve and the contact elements at the inner side get more pressed, see Figures 23a and 23b. The corresponding contact situation in this stationary cornering situation is depicted in Figure 23c. The distortion of the footprint as a result of cornering is clearly observed.

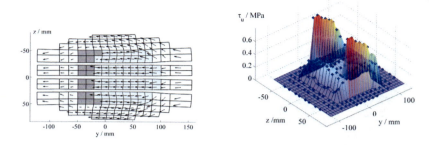

Fig. 22 Tractive rolling tire (braking slip of 8%.)

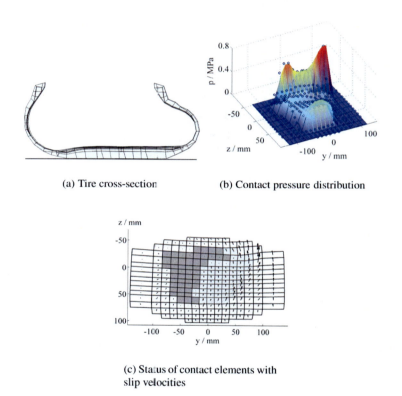

(a) Tire cross-section (b) Contact pressure distribution

(c) Status of contact elements with slip velocities

Fig. 23 Cornering effects for a slip angle $\delta = 3°$.

7 Remark to the Computational Effort

The proposed algorithm has been coded in MATLAB programming language mainly, where only computational expensive standard routines, like finite element subroutines, etc., have been translated into FORTRAN code and embedded as link libraries. In addition, external libraries, for example the parallel direct PARDISO solver [29] have been utilized for the solution of the non-symmetric linear systems. This MATLAB based system provides a powerful and efficient environment for the development of innovative technologies in computational mechanics, for example with respect to a flexible treatment of data structures, etc. Besides the afore mentioned techniques to speed up the computations a great potential regarding runtime optimization for special purpose applications remains open. The statistical data outlined below therefore should be judged as upper limit at the present state of the art of computer science technology.

The overall computation time for the free rolling state of the tire model presented in the previous section took about 15 minutes on a notebook computer including an INTEL T2400 Duo-processor and 2 Gb RAM. Additional steps for tractive rolling simulations like braking or cornering are easily performed using restart option with a few minutes

If, in addition, inelastic constitutive properties are taken into account, the computation time increases significantly. Nevertheless, the methods introduced in Section 4 provide a mathematical reliable basis with great potential for further optimization with respect to computational efficiency.

8 Summery and Conclusions

The relative kinematic description for rolling contact analysis within the Arbitrary Lagrangian Eulerian framework has been accepted as well established standard for finite element computations. A major disadvantage is, that history dependent effects are not treatable by standard techniques known from computational solid body mechanics. This contribution focussed on two key problems, namely the treatment of inelastic constitutive behavior within the spatial ALE observer framework and the numerical efficient treatment of the frictional lateral contact constraints for stationary rolling simulations.

With special focus on applications to rubber wheels a brief outline on the constitutive modeling of the rather complicated behavior of particle filled rubber material has been presented. A hierarchical modelling approach based on phenomenological observations is described in Section 3 within a thermodynamic consistent continuums mechanics framework. With focus on computability the established concept of internal variables for dissipative material behavior has been recapitulated. The implemented material library contains options for the computational treatment of incompressibility constraints, a hierarchical class of hyperelastic models, a frequency domain bridging approach for the description of viscoelastic properties of elastomers and alternative methods for damage effects. These developments are underlined by comparison of computational results with experiments performed with well defined rubber composites.

Special attention has been paid on the computational treatment of the corresponding evolution equations for the inelastic internal variables within the ALE framework. With emphasis to the numerical effort focussing onto industrial applications an explicit scheme based on the so called fractional step method has been suggested. In a first step well established schemes for the solution of the local evolution of inelastic variables can be adapted immediately. With regard to the second step, the advection of inelastic variables due to the particle motion within the spatially fixed finite element mesh, only unsatisfying numerical solutions have been evaluated from literature. In this presentation we clearly figured out the advantage of Time-Discontinuous-Galerkin (TDG) methods for the numerical solution of advection dominated problems. There are no restriction with respect to critical time step size and the accuracy of the scheme can be easily increased by simply increasing the order of the ansatz-polynomials for the temporal approximation. Furthermore, it is easily implemented are there is no need for structured spatial meshes. The physical consistency of the computational approach has been demonstrated, while all computations have been performed in 3D. The computational effort increases mentionable in comparison to a pure elastic computation, because in each evolution step an additional non-symmetric system for the advection step has to be solved, whose size increases with the ansatz-order of the temporal approximation. Never the less, this approach is based on a sound mathematical formulation, which enables for adaptive error control with respect to the temporal as well as the spatial discretization. Topics like increased numerical efficiency, adaptive error control, etc., are subjects for further research to provide reliable and efficient numerical methods for industrial applications.

A second aspect of this report is on the efficient and stable treatment of frictional rolling contact problems. Here the focus has been led on a physical reliable solution of the stick conditions within the ALE framework. Based on the idea of a material description of particles in contact moving through the finite contact area, a mixed interpolation scheme has been introduced to compute simultaneously both, the spatial deformation of the contact nodes and the relative movement (slip) of the contacting material particles. This yields a fully implicit algorithm for the enforcement of stick conditions. Because the computed slip variables provide a material representation of the lateral contact, well established methods for frictional contact within Lagrangian mechanics are adapted easily. It has been demonstrated, that for each step of the computation leads to quadratic convergence within the Newton–Raphson scheme. The physical consistency as well as the industrial applicability has been demonstrated by computational studies with three-dimensional models for Grosch wheels and quite detailed tire models. The effort for these computations is manageable on modern notebook computers within acceptable time.

This contribution marks the current state of the art on computational methods for rolling rubber wheels. However, especially regarding tire mechanics application there is an open field on future research. Actually we are far away to state that computational mechanics can predict overall tire properties from scratch. Essential points on this way have been touched within the Research Unit 492, like the investigation of rubber road contact using micro-mechanical approaches. But a convergence of more traditional statistically based models and computational multi-scale approaches has not been obtained. We estimate, that a mayor reason for this lays in the insufficient knowledge on the thermomechanical behavior of rubber and its proper description for the different model approaches. Yet now, a broad variety of phenomenological rubber models is available in commercial finite element codes, which provide reasonable resuts for specific test conditions with well investigated constitutive parameters.

Acknowledgements

The authors express their gratitude for the the financial support by the German Research Foundation (DFG) and the fruitful collaboration within the framework of the DFG-Research Unit 492.

References

1. Baines, M.J.: A survey of numerical schemes for advection, the shallow water equations and dambreak problems. In: Proceedings of the 1st CADAM Workshop, Wallingford (1998)
2. Bauer, R.E.: Discontinuous Galerkin methods for ordinary differential equations, Master's Thesis, University of Northern Colorado (1995)
3. Bayoumi, H.N., Gadala, M.S.: A complete finite element treatment for the fully coupled implicit ALE formulation. Computational Mechanics 33, 435–452 (2004)
4. Benson, D.: An efficient accurate simple ALE method for nonlinear finite element programs. Computer Methods in Applied Mechanical Engineering 72, 305–350 (1989)
5. Benson, D.: Computational methods in Lagrangian and Eulerian hydrocodes. Computer Methods in Applied Mechanical Engineering 99, 235–394 (1992)
6. Blab, R., Harvey, J.T.: Modeling measured 3d tire contact stresses in a viscoelastic FE-pavement model. The International Journal of Geomechanics 2(3), 271–290 (2002)
7. Carter, F.W.: On the action of a locomotive driving wheel. Proceedings of the Royal Society London A 122, 151–157 (1926)
8. Cockburn, B., Karniadakis, G.E., Shu, C.W.: The development of discontinuous Galerkin methods. In: Cockburn, B., Karniadakis, G.E., Shu, W. (eds.) Discontinuous Galerkin Methods, pp. 3–50. Springer, Heidelberg (2000)
9. Davy, F.: Modelling and numerical simulation of filled rubber, Master's Thesis, Leibniz Universität Hannover (2006)

10. Donea, J., Huerta, A., Ponthot, J.P., Rodriguez-Ferran, A.: Arbitrary Lagrangian–Eulerian methods. In: Stein, E., de Borst, R., Hughes, T. (eds.) Encyclopedia of Computational Mechanics. Fundamentals, vol. 1, pp. 414–437. John Wiley & Sons, Chichester (2004)
11. Faria, L.O., Oden, J.T., Yavari, B., Tworzydlo, W., Bass, J.M., Becker, E.B.: Tire modeling by finite elements. Tire Science & Technology 20, 33–56 (1992)
12. Fressmann, D., Wriggers, P.: Advection approaches for single- and multi-material arbitrary Lagrangian–Eulerian finite element procedures. Computational Mechanics 39, 153–190 (2005)
13. Gadala, M.S.: Recent trends in ALE-formulation and its applications in solid mechanics. Computer Methods in Applied Mechanical Engineering 193, 4247–4275 (2004)
14. Godunov, S.: Finite difference method for numerical computation of discontinuous solutions of the equation of fluid dynamics. Math. Sbornik 47, 272–306 (1959)
15. Guo, Z., Sluys, L.J.: Computational modelling of the stress-softening phenomenon of rubber-like materials under cyclic loading. European Journal of Mechanics A/Solids 25, 877–896 (2006)
16. Hartmann, S.: Finite-elemente Berechnung inelastischer Kontinua. Universität Kassel, Habilitation (2003)
17. Heinrich, G., Kaliske, M.: Theoretical and numerical formulation of a molecular based constitutive tube-model of rubber elasticity. Computational and Theoretical Polymer Science 7, 227–241 (1998)
18. Hu, G., Wriggers, P.: On the adaptive finite element model of steady-state rolling contact for hyperelasticity in finite deformations. Computer Methods in Applied Mechanics and Engineering 191, 1333–1348 (2002)
19. Hughes, T.J.R., Hulbert, G.M.: Space-time finite element methods for elastodynamics. Computer Methods in Applied Mechanical Engineering 66, 339–363 (1988)
20. Koehne, S.H., Matute, B., Mundl, R.: Evaluation of tire tread and body interactions in the contact patch. Tire Science & Technology 31(3), 159–172 (2003)
21. Kolditz, O.: Computational Methods in Environmental Fluid Mechanics. Springer, Heidelberg (2002)
22. Laursen, T.A.: Computational Contact and Impact Mechanics. Springer, Heidelberg (2002)
23. Le Tallec, P., Rahier, C.: Numerical models of steady rolling for non-linear viscoelastic structures in finite deformations. International Journal on Numerical Methods in Engineering 37, 1159–1186 (1994)
24. Nackenhorst, U.: The ALE-formulation of bodies in rolling contact – Theoretical foundations and finite element approach. Computer Methods in Applied Mechanical Engineering 193(39-41), 4299–4322 (2004)
25. Nasdala, L., Kaliske, M., Becker, A., Rothert, H.: An efficient viscoelastic formulation for steady-state rolling. Computational Mechanics 22, 395–403 (1998)
26. Oden, J.T., Lin, T.L.: On the general rolling contact problem for finite deformations of a viscoelastic cylinder. Computer Methods in Applied Mechanical Engineering 57, 297–367 (1986)
27. Ogden, R.W., Roxburgh, D.G.: A pseudo-elastic model for the Mullins effect in filled rubber. Proceedings of the Royal Society London A 455, 459–490 (1999)

28. Rodriguez-Ferran, A., Casadei, F., Huerta, A.: ALE stress update for transient and quasistatic processes. International Journal on Numerical Methods in Engineering 43, 241–262 (1998)
29. Schenk, O., Gärtner, K.: Solving unsymmetric sparse systems of linear equations with Pardiso. Journal of Future Generation Computer Systems 20(3), 475–487 (2004)
30. Shakib, F., Hughes, T.J.R.: A new finite element formulation for computational fluid dynamics: IX. Fourier analysis of space-time Galerkin/least-squares algorithms. Computer Methods in Applied Mechanics and Engineering 87, 35–58 (1991)
31. Simo, J.C.: On a fully three-dimensional finite-strain viscoelastic damage model. Computer Methods in Applied Mechanical Engineering 60, 153–173 (1987)
32. Simo, J.C., Hughes, T.J.R.: Computational Inelasticity. Springer, Heidelberg (1998)
33. Stanciulescu, I., Laursen, T.: On the interaction of frictional formulations with bifurcation phenomena in hyperelastic steady state rolling calculations. International Journal of Solids and Structures 43, 2959–2988 (2006)
34. Stoker, C.: Developments of the arbitrary Lagrangian–Eulerian method in nonlinear solid mechanics, PhD Thesis, University Twente (1999)
35. Wriggers, P.: Finite element algorithms for contact problems. Archives of Computer Methods in Engineering 2, 1–49 (1995)
36. Wriggers, P.: Computational Contact Mechanics, 2nd edn. Springer, Heidelberg (2006)
37. Ziefle, M.: Numerische Konzepte zur Behandlung inelastischer Effekte beim reibungsbehafteten Rollkontakt, PhD Thesis, Leibniz Universität, Hannover (2007)
38. Ziefle, M., Nackenhorst, U.: An internal variable update proceedure for the treatment of inelastic material behavior within an ALE-description of rolling contact. Applied Mechanics and Materials 9, 157–171 (2008)
39. Ziefle, M., Nackenhorst, U.: Numerical techniques for rolling rubber wheels – Treatment of inelastic material properties and frictional contact. Computational Mechanics 43, 337–356 (2008)

Simulation and Experimental Investigations of the Dynamic Interaction between Tyre Tread Block and Road

Patrick Moldenhauer and Matthias Kröger

Abstract. The dynamic behaviour of a tyre tread block is described by a modularly arranged model with a high numerical efficiency. Special emphasis is laid on the interaction between structural dynamics and contact mechanics. Parameter-dependent friction is implemented as well as a non-linear contact stiffness to consider the properties of the rough surface. The model includes local wear which influences the dynamic behaviour. The relevant parameters are obtained by experimental investigations. High-frequency stick-slip vibrations of the tread block occur e.g. on a corundum surface within a certain parameter range which are compared to simulation results. Furthermore, the model includes the rolling process of the tyre: the tread block follows a trajectory which is obtained from the deformation of the tyre belt. The results from simulations with a single tread block provide a deeper insight into highly dynamic processes that occur in the contact patch such as tyre squeal, run-in or snap-out effects.

1 Introduction

Tyre tread blocks are permanently subjected to different driving states such as acceleration or ABS braking as well as cornering. Even for a constant vehicle velocity the tread blocks experience an excitation resulting from the tyre deformation in the contact patch and from friction induced vibrations. The contact time of a tread block for a vehicle velocity of 100 km/h takes about 5 ms. During this contact the tread block passes a sticking and a sliding phase and finally snaps out. This leads to an unsteady dynamic behaviour

---

Patrick Moldenhauer · Matthias Kröger
Institute of Machine Elements, Engineering Design and Manufacturing,
Technical University Freiberg, Agricolastr. 1, 09596 Freiberg, Germany
e-mail: {patrick.moldenhauer, kroeger}@imkf.tu-freiberg.de

of the tread blocks which has to be covered in models to achieve a realistic simulation. On the one hand, present tread block models usually show a high grade of complexity. For example, in [17] a three-dimensional finite element approach with non-linear material description and thermo-mechanical coupling is presented. Such models are only applicable within limits for dynamic calculations because of the computational effort. On the other hand there are models that treat the tread block as a simple elastic beam [5] or time-dependent viscoelastic springs [28]. Here, the block geometry or important structural effects are neglected. The aim of this approach is to compromise between these concepts to calculate dynamic processes and vibrations of tyre tread blocks. This means to model the geometric and dynamic properties of the tread block and their interaction with the road surface including the frictional contact as well as the rolling contact without the loss of numerical efficiency.

2 Modular Tread Block Model

In many of applications the frictional contact effects technical problems. This is especially the case for contacts with at least one component made of elastomers and if relative motion occurs between the contact partners. Typical unwanted vibration effects are juddering windscreen wipers or side window seals of cars, stick-slip oscillations of seals and the noise from squealing tyres. Elastomer contacts are difficult to describe, because a large number of influencing parameters must be considered when modelling these contacts, see Figure 1.

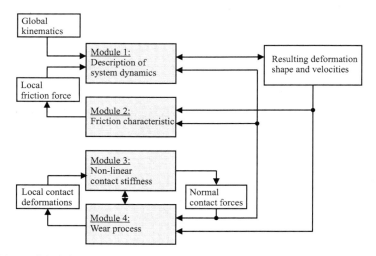

Fig. 1 Modular modelling of dynamic systems with elastomer contacts.

Dynamic Interaction between Tyre Tread Block and Road

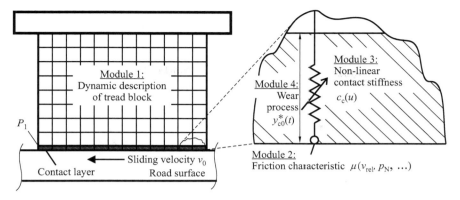

Fig. 2 Dynamic tyre tread block model.

The following section describes the concept for the dynamic tread block modelling. The basic idea is to model the structural dynamics and the contact mechanics at first separately and to link them during the simulation in a reasonable way. Point contact elements P_i describe the local contact phenomena and are coupled to the structural model, see Figure 2. The structural model provides the input for the point contact elements e.g. the present relative velocity v_{rel} between point contact element and road surface. The point contact elements with their lengths $y_{c0}^*(t)$ form the contact layer and calculate the local contact forces which in turn are applied to the block layer as external forces.

Each mechanical or physical effect is described by one module whereas the single modules are linked and interact with each other during the simulation. For the description of the structural dynamics it is essential to include the tread block dimensions as well as the mechanically relevant material properties. Therefore, a finite element model is generated in advance which represents the tread block structure. The contact layer with its point contact elements considers parameter-dependent friction by local friction forces. Furthermore, the rough road surface effects a non-linear relation between normal force and normal displacement. This effect is covered by a non-linear contact stiffness $c_c(u)$ which is implemented in the point contact elements. The contact forces are then applied to the structural model in each time step. Moreover, the model considers parameter-dependent wear effects within the point contact elements. The overall model can, therefore, be divided into the following modules which are explained in the following sections:

- Module 1: Dynamic tread block description,
- Module 2: Local friction characteristic,
- Module 3: Non-linear contact stiffness,
- Module 4: Wear.

Fig. 3 Summer tyre (left) and winter tyre (right). Source: Continental AG, Hannover, Germany.

2.1 Module 1: Dynamic Tread Block Description

This module considers dynamic structural effects from inertia, elasticity and damping. One input parameter of this module is the tread block geometry. Different vehicle tyres show a variety of block shapes and profile heights. New winter tyres for passenger cars provide a tread depth of about 8 mm whereas the legal limit for the profile depth in the European Union is 1.6 mm. With respect to their function tread elements in the shoulder region are mostly designed as blocks whereas in the centre increasingly band elements are used, see Figure 3.

Depending on the tread block geometry a certain mechanical behaviour can be observed. Long and flat tread blocks lead to a stiff behaviour in circumferential direction. They are often used for summer tyres. On the contrary, a winter tread is characterised by siped tread blocks which result in a comparably small geometric stiffness. In addition, the tyre properties can be optimised by the compound with regard to material stiffness and damping [48].

The presented model considers the tread block geometry by a finite element model which has to be generated in advance. In this work a linear material model is used because a numerical efficient modelling is required which simulates highly dynamic processes in the kHz regime. This demands a reduction of the degrees of freedom which will be discussed later on. Basically, it is possible to employ a non-linear material model, but standard methods for the reduction of degrees of freedom can then not be applied.

The linear finite element model requires as input parameters the tread block geometry with the corresponding discretisation, the elasticity modulus E, the density ρ, Poisson's ratio ν and for the 2D case the state of stress. The corresponding mass matrix $\mathbf{M}$ and the stiffness matrix $\mathbf{K}$ is extracted from the mesh generating tool to the program MATLAB in which the presented model is built up. In the following, Rayleigh damping is assumed which leads to a damping matrix $\mathbf{D} = \alpha \mathbf{M} + \beta \mathbf{K}$ with the scalar coefficients α and β. Furthermore, the potential contact nodes are defined where the contact forces

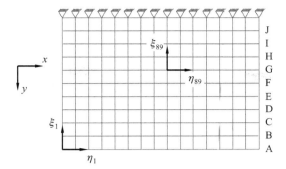

Fig. 4 Numbering of finite element nodes.

of the single point contact elements are applied. Here, these are the nodes at the tread block bottom.

The nodes of the plane finite element model are counted line-by-line from bottom left to top right. The horizontal and vertical displacements η_i, ξ_i of the nodes are exemplarily shown in Figure 4 for $i = 1$ and $i = 89$. The global frame with x and y direction for the following investigations is depicted as well. The nodes at the right edge are additionally denoted with A to J to describe these nodes later on with respect to simulation results.

With respect to the high numerical efficiency the structural model has to be reduced with regard to its degrees of freedom. The presented model employs the Craig–Bampton reduction method which can be interpreted as the combination of static and modal condensation [7]. However, it is also possible to use other reduction methods e.g. pure modal condensation, pure static condensation, Guyan or Martinez reduction, or Krylov subspace method [8, 40, 41, 45]. The advantage of the Craig–Bampton method is, that certain degrees of freedom which have to be defined in advance, are explicitly retained in the reduced system. Here, these are the contact nodes. Therefore, contact algorithms which have been developed for the finite element method can be used without limitation.

The Craig–Bampton reduction always fulfils the static solution of the system and the additional modal ansatz functions approximate the dynamic behaviour. The number of additional modes determines the approximation quality: a larger number of modal ansatz functions improves the dynamic description but leads to a higher numerical effort.

Nearly all reduction methods are based on the superposition principle and assume a linear structural system which is given here by the linear material model. However, there are reduction methods for non-linear material properties but the actual numerical effort for those systems can be compared to the non-reduced systems [3, 35]. In general, it is not compulsory for the tread block model to employ a reduction of the degrees of freedom. All other modules also work with a full finite element description and a non-linear material description.

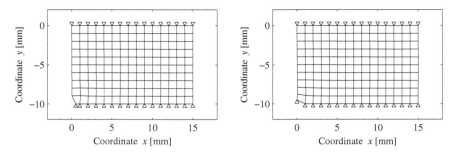

Fig. 5 Constraint modes for the node at bottom left in horizontal and vertical direction.

The Craig–Bampton reduction allows the retention of physically important degrees of freedom as primary degrees of freedoms. All other degrees of freedom are described as secondary degrees of freedom and are modally condensed. The tread block system here can be interpreted as a linear system with non-linear couplings which are the single point contact elements. The primary degrees of freedom are expressed by shape functions which provide a unit displacement at the respective degree of freedom while all other primary degrees of freedom are constrained. The corresponding displacements of the secondary degrees of freedom result from the static relation. These shape functions are called constraint modes. Figure 5 exemplarily depicts the constraint mode for the bottom left node in horizontal and vertical direction ξ_1, η_1.

Relative displacements between the secondary and the primary degrees of freedom which result from dynamic effects are approximated by additional shape functions which are named normal modes. These normal mode shape functions have to be fixed at the primary degrees of freedom to explicitly retain the primary degrees of freedom. It is a reasonable but not necessary choice to use as normal modes the modal shape functions of the system with fixed primary degrees of freedom. Figure 6 shows the first six normal modes for the tread block.

The Craig–Bampton transformation is exact if the constraint modes and all normal modes of the system are considered. The reduction of the degrees of freedom reveals if in addition to the constraint modes only some important normal modes are used. The corresponding reduced system matrices are generated by a left and right multiplication with the Craig–Bampton transformation matrix $\mathbf{T}_{\text{red}}$,

$$\mathbf{M}_{\text{red}} = \mathbf{T}_{\text{red}}^{\text{T}} \mathbf{M} \mathbf{T}_{\text{red}}, \quad \mathbf{D}_{\text{red}} = \mathbf{T}_{\text{red}}^{\text{T}} \mathbf{D} \mathbf{T}_{\text{red}}, \quad \mathbf{K}_{\text{red}} = \mathbf{T}_{\text{red}}^{\text{T}} \mathbf{K} \mathbf{T}_{\text{red}}. \quad (1)$$

External forces $\mathbf{F}_{\text{E}}$ which can only be applied at primary degrees of freedom are projected into the Craig–Bampton description as well by the transformation matrix $\mathbf{T}_{\text{red}}$

$$\mathbf{F}_{\text{red}} = \mathbf{T}_{\text{red}}^{\text{T}} \mathbf{F}_{\text{E}}. \quad (2)$$

Dynamic Interaction between Tyre Tread Block and Road

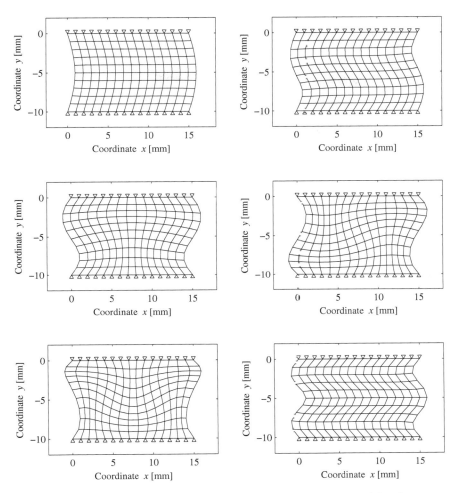

Fig. 6 First six normal modes with fixed primary degrees of freedom.

This leads to the transformed system equation with the reduced displacement vector x_{red}

$$M_{red}\ddot{x}_{red} + D_{red}\dot{x}_{red} + K_{red}x_{red} = F_{red}. \qquad (3)$$

The dynamic response of the primary degrees of freedom is directly calculated with this reduced description. If necessary, the response of the secondary degrees of freedom can be calculated after the simulation by a backward transformation. Depending on the numerical time integration step, the simulation time and the dimension of the displacement vector x_{red} the backward transformation may last as long as the simulation itself.

2.2 Module 2: Local Friction Characteristic

Rubber friction mechanisms can be divided according to their physical origin. The classification of Kummer from 1966 still holds today where four components are proposed as unified theory of rubber friction which are briefly explained in the following [25].

Hysteresis Component

Rubber shows a significant internal material damping. If the rubber undergoes a deformation only a part of the applied work of deformation can be regained when a rubber sample slides on a rough surface. The dissipated energy results in a friction force. This hysteretic material property can be described with a viscoelastic material model. With respect to the relative velocity the dissipated energy shows a characteristic maximum. Hysteresis friction is assumed to be the dominating contribution of the overall friction force on wet rough surfaces [14, 16].

Adhesion Component

Molecular attraction forces are generated by the direct contact between rubber and contact partner. The magnitude of the adhesion effect strongly depends on the type of rubber compound, the contact partner and filler particles. The adhesion component dominates on smooth dry surfaces [1, 22, 26].

Cohesion Component

The generation of cracks and wear in the rubber material needs additional energy which results in the cohesion friction force [6, 20].

Viscous Friction

Viscous friction occurs if there is a fluid film between the contact partners. Due to the viscosity of the fluid shear forces are generated that are directed against the sliding direction [47].

For the description of rubber friction many approaches have been developed which can be divided in different categories:

Analytical approaches are usually based on single rubber elements that slide over a rough surface [26, 29, 37]. The rubber element considers the stiffness and damping properties by a linear one-dimensional description. The models either work in the time or frequency domain and are appropriate to perform fundamental studies while structural effects cannot be covered by the 1D approach.

Numerical approaches are usually realised by a finite element calculation. Here, the geometry of the rubber component can be considered in detail and

Dynamic Interaction between Tyre Tread Block and Road   173

also thermal investigations can be carried out [17, 52]. Recent models deal with crack generation and propagation [6, 20]. These models need assumptions about the frictional behaviour of the contact partners to predict the mechanical behaviour of the entire system. Finite element models are characterised by a high numerical effort. To overcome this problem multi-scale approaches or reduced models are used [15, 34].

The experimental approach is very pragmatic and often useful because all physical effects are measured with their interactions. With a certain experimental configuration it is possible to enforce or suppress e.g. hysteresis or adhesion friction [27, 32]. However, experiments are time and cost consuming and are only valid for the tested configurations. Furthermore, the friction process on rough surfaces reveals a stochastic character. The analysis of statistical parameters of the friction coefficient values shows e.g. a correlation to a Gaussian distribution [11].

The tread block model requires as one input the local friction characteristic in order to calculate the system dynamics for the whole tread block. Presently, the model considers the dependence of the friction coefficient on the relative velocity and normal pressure. This friction characteristic has to be determined in advance as a result of an analytical, numerical or experimental approach. The implementation into the model can either be done by an analytical function, as a result of a local friction coefficient simulation, or as an approximation function of experimental data. Furthermore, experimental data points can directly be included within the simulation. However, this approach is not reasonable from a physical point of view: at the sampling points the friction characteristic is in the general case not continuous which leads to mistakable dynamic effects. During the simulation routine the tread block model provides the input parameters for the friction characteristic: the relative velocity and the normal pressure are calculated in every time step at each point contact element. The friction coefficient is multiplied with the respective normal contact force and is applied as friction force at the respective point contact element. In the sliding state the friction force is directed against the sliding direction, cf. [42]. This behaviour can be expressed by the sign function

$$F_R = \text{sgn}(v_{\text{rel}}) \, \mu(v_{\text{rel}}, p_N, \ldots) F_N. \qquad (4)$$

This leads to a non-smooth friction characteristic due to the step at $v_{\text{rel}} = 0$, cf. Figure 7, which requires a distinction of cases in the model. From a numerical point of view a smoothing of this characteristic is useful e.g. by an approximation of the sign function by an arc tangent function

$$F_R = \frac{2}{\pi} \arctan(k_S \, v_{\text{rel}}) \, \mu(v_{\text{rel}}, p_N, \ldots) F_N. \qquad (5)$$

The advantage of this approach is that no case distinction between $v_{\text{rel}} < 0$ and $v_{\text{rel}} > 0$ is necessary. However, an exact sticking condition cannot be covered by this approximation, since $\mu(v_{\text{rel}} = 0) = 0$. The slope at $v_{\text{rel}} = 0$

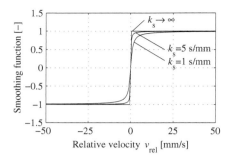

Fig. 7 Numerical approximation of the sign function under variation of slope parameter k_S.

determines the quality of the approximation and is, therefore, governed by the parameter k_S. The larger this value is chosen the better is the approximation of the sign function which leads at the same time to a stiffer system behaviour. Consequently, smaller simulation time steps increase the computational time. Figure 7 depicts exemplarily the influence of the slope parameter k_s on the smoothing function.

Rubber friction characteristics usually show a maximum with respect to the sliding velocity v_{rel}. Therefore, an exponentially decreasing approximation function has been chosen with the coefficients $\mu_{\infty,v}$ $\mu_{0,v}$ and γ_v. The multiplication with the smoothing function generates the characteristic maximum. Furthermore, a decreasing friction coefficient with increasing normal pressure p_N is observed in nearly all experiments. Therefore, an exponentially decreasing approximation function with the coefficients $\mu_{0,p}$, $\mu_{\infty,p}$ and γ_p has been applied. This leads to an overall approximation function which is capable of covering a typical rubber friction characteristic

$$\mu(v_{\text{rel}}) = \frac{2}{\pi} \arctan\left(k_S\, v_{\text{rel}}\right) \left(\mu_{\infty,v} + (\mu_{0,v} - \mu_{\infty,v})\, e^{(\gamma_v\, |v_{\text{rel}}|)}\right)$$
$$\cdot \left(\mu_{\infty,p} + (\mu_{0,p} - \mu_{\infty,p})\, e^{(\gamma_p\, p_N)}\right). \tag{6}$$

Experimentally identified coefficients for two different rough surfaces are given in Section 3.4. In general, other approximation functions can be implemented in the model as well.

2.3 Module 3: Non-linear Contact Stiffness

If two bodies contact each other whereas at least one body has a rough surface a plane contact area cannot be assumed. At first, the bodies contact at the highest asperities and provide single contact points [12, 13, 37]. In case of the tyre/road contact it is reasonable to model the tyre tread compound as elastic

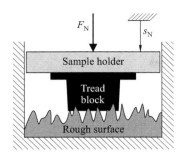 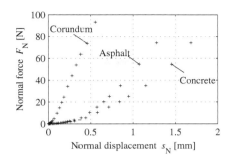

Fig. 8 Left: Experimental setup for normal contact stiffness investigations. Right: Measured normal force-displacement characteristic [10].

and the road surface as rigid. This assumption is justified by the large difference of material stiffness of the contact partners. With increasing normal contact pressure more and more contact points appear due to the elastic tread material which results in an increase of the real contact area [12].

This effect leads to a non-linear relation between normal force and normal displacement: at very small contact forces there are only a few contact points which result in high local deformations. This effects a small resistance against penetration. At higher normal forces the load distributes over a large number of contacting points leading to a higher penetration resistance. Experiments from Gäbel with a tread block on rough surfaces exemplarily show the relation between normal force and normal displacement for different rough surfaces, see Figure 8. All curves provide a similar shape with a small slope at small penetration depths. Above a certain normal displacement s_N, which depends on the respective surface properties, an approximately linear behaviour is observed [24].

The resistance against penetration is defined as the global normal contact stiffness c_N and is calculated by the ratio of normal force increment ΔF_N and normal displacement increment Δs_N

$$c_N = \frac{\Delta F_N}{\Delta s_N}. \qquad (7)$$

With increasing penetration depth a saturation effect can occur and the global contact stiffness then reaches the structural stiffness. Due to the large number of contact points the contact situation is similar to a plane contact. This leads to a linear relation between normal force and normal displacement in this regime and, therefore, to a constant contact stiffness. This is the case for the corundum surface in Figure 8 (right) but not for the road surfaces asphalt and concrete for realistic normal tyre loads.

The tread block model has to consider this effect without losing numerical efficiency. The classical approach is to compute the contact with the

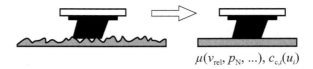

Fig. 9 Numerically efficient modelling of the rough contact as smooth with local friction characteristic and non-linear contact stiffness.

discretised rough road surface. One numerical contact algorithm is the penalty method. It is well-known in the field of finite element simulations [49]. This approach admits a penetration of the contacting bodies. However, this penetration is penalised by the so-called penalty-stiffness. The resulting separating force is proportional to the amount of penetration and represents the normal contact force. The higher the penalty stiffness is defined with regard to the structural stiffness the less is the resulting penetration. However, a high penalty factor leads to a stiffer system which in turn leads generally to a higher numerical effort. Furthermore, the penalty stiffness has no physical meaning.

However, this contact algorithm is numerically expensive on discretised rough surfaces and, therefore, does not comply with the above mentioned requirements. The approach presented here is similar to the penalty method: the road surface is modelled as a smooth surface which leads to a fast and efficient contact algorithm, see Figure 9. The above discussed non-linear relation between normal force and displacement is considered by non-linear springs in the point contact elements. The point contact elements which are coupled to the block layer deliver the local normal contact forces. In contrast to the penalty spring elements with a constant stiffness now non-linear contact stiffness elements are applied. They consider the above described non-linear relation between normal force and displacement which originates from the contact of the rubber with the rough road surface and cannot be considered by the linear material model of the block layer. Compared to the original penalty method the non-linear contact stiffness leads to a larger penetration of the contacting bodies. However, this larger penetration makes sense from a contact mechanical point of view. The penetration of the bodies which are both modelled as smooth represent the gradual penetration of the rough road surface into the soft rubber material. Figure 10 exemplarily depicts the reference lengths and the coordinates which are relevant for the contact algorithm for the ith point contact P_i.

The initial uncompressed length of the non-linear spring is y_{c0}. The vertical distance from the fixed support to the ith contact node denotes $y_{A,i}(t)$. The length $y_{A,i}(t)$ depends on the vertical displacement $\eta_i(t)$ of the ith node and results from the structural model during the simulation at each time step

$$y_{A,i}(t) = H(t) - \eta_i(t). \tag{8}$$

Dynamic Interaction between Tyre Tread Block and Road 177

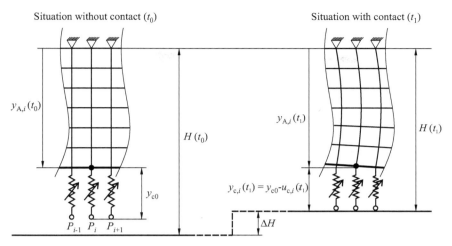

Fig. 10 Reference lengths and coordinates for the contact algorithm.

The length $H(t)$ describes the vertical distance between the fixed support and the surface on which the tread block slides. Therefore, the length $H(t)$ defines the compression of the whole tread block in vertical direction. A certain normal load can implicitly be applied by the adjustment of this parameter. The length $H(t)$ has to be defined in advance and can be formulated, in general, time dependent.

The contact algorithm works as follows: if the sum of the current block layer length $y_{A,i}(t)$ and the length of the undeformed non-linear spring y_{c0} is less than the compression preset $H(t)$ no contact occurs at the ith point contact element and the local normal force $F_{N,i}(t)$ as well as the local friction force $F_{R,i}(t)$ are set to zero, see Figure 10 left. In the other case contact is detected at the respective point contact element, see Figure 10 (right). Then, the local normal force $F_{N,i}(t)$ is calculated by the non-linear contact stiffness $c_{c,i}(u_i(t))$ and the present compression $u_i(t)$ of the ith spring. With respect to readability, the time-dependent notation is omitted in the following:

$$F_{N,i} = \begin{cases} 0 & \text{if } y_{A,i} + y_{c0} < H, \\ c_{c,i}(u_i) u_i & \text{else.} \end{cases} \quad (9)$$

The compression u_i of the non-linear spring at the ith point contact element is calculated with the following algorithm:

$$u_i = y_{c0} - y_{c,i} \quad (10)$$

$$= y_{c0} - (H - y_{A,i}) \quad (11)$$

$$= -(H - (y_{A,i} + y_{c0})). \quad (12)$$

The local friction force $F_{\mathrm{R},i}$ at the ith point contact element is determined by the local friction coefficient $\mu_i(v_{\mathrm{rel},i}, p_{\mathrm{N},i})$ and the local normal force $F_{\mathrm{N},i}$

$$F_{\mathrm{R},i} = \mu_i\left(v_{\mathrm{rel},i}, p_{\mathrm{N},i}\right) F_{\mathrm{N},i}. \tag{13}$$

The non-linear contact stiffness is assumed to obey an exponential relation between normal force and displacement, cf. [24, 38]

$$c_{\mathrm{c},i}(u_i) = c_\infty (1 - \mathrm{e}^{-k u_i}). \tag{14}$$

The coefficients c_∞ and k have to be identified with respect to the contact partners e.g. from experiments which is discussed in Section 3.5. Basically, other laws for the relation between normal force and normal displacement can be implemented as well.

2.4 Module 4: Wear

Rubber materials show a comparably low wear resistance in contrast to other classical engineering materials such as metals. The wear of tyre treads is determined by many influencing parameters: on the one hand there are different construction parameters e.g. tread design, tyre body contour or contact patch formation with corresponding contact pressure distribution. On the other hand wear depends on ambient conditions like driving style, ambient temperature or road properties. In practice there are outdoor tests of vehicle tyres which are time and cost consuming. A typical wear test lasts 20,000 to 40,000 km and several weeks up to some months. Indoor tests e.g. on drum testers provide a much cheaper alternative. Due to the severe testing conditions a shorter time is required to achieve a wear level which is similar to the outdoor tests. However, the correlation between both test methods is not always satisfactory [44, 51]. The most effective method with respect to costs and time are wear predictions by simulation techniques. Here, no test tyre needs to be produced, test stand time does not apply and there is no testing data to be analysed.

To consider the wear process in the tread block model it is necessary to find an adequate wear law. In the literature there is a number of equations for the wear calculation which are mostly referred in a way to the frictional power and mostly describe the mass loss per friction time $\dot{m} = \mathrm{d}m/\mathrm{d}t$ or sliding distance $\eta_{\mathrm{wear}} = \mathrm{d}m/\mathrm{d}s$, see [2, 9, 17, 50] or [46]. If the velocity v is known both notations $\dot{m}$ and η_{wear} can be transferred

$$\dot{m} = \frac{\mathrm{d}m}{\mathrm{d}t} = \frac{\mathrm{d}m}{\mathrm{d}s}\frac{\mathrm{d}s}{\mathrm{d}t} = \frac{\mathrm{d}m}{\mathrm{d}s} v = \eta_{\mathrm{wear}}\, v. \tag{15}$$

According to these laws the wear rate is related in general to the local pressure p_N, velocity v and the friction coefficient μ. Depending on the respective wear

Dynamic Interaction between Tyre Tread Block and Road   179

laws in the literature there are coefficients b_i to consider non-linear wear behaviour. A generalised wear law can be written as

$$\dot{m} = b_0 \mu^{b_1} p_N^{b_2} v_{\text{rel}}^{b_3}. \tag{16}$$

To implement wear in the tread block model the non-linear springs that consider the non-linear contact stiffness within the point contact elements can decrease their initial lengths according to a wear law. Assuming a homogeneous material the mass loss per time can be transferred to a decreasing length of the non-linear springs $\ell_{w,i}(t)$ by the density ρ and the corresponding point contact area A_i

$$\dot{\ell}_{w,i}(t) = \frac{1}{\rho A_i} \dot{m}_i(t). \tag{17}$$

The remaining length of the non-linear springs $y^*_{c0,i}$ is calculated during the simulation by time integration of Equation (17) at each point contact

$$y^*_{c0,i}(t) = y_{c0} - \int \dot{\ell}_{w,i}(t) \, dt. \tag{18}$$

As a first approach the law of Fleischer ($b_1 = b_2 = b_3 = 1$ in Equation (16)) is used for the wear calculation in the module [9].

3 Parameter Identification

As a next step the parameter identification for the single modules is carried out. The application of the Craig–Bampton reduction method requires the material characterisation by means of the mass, stiffness and damping matrix. These are generated by a finite element meshing tool which needs the tread block geometry and material properties as input parameters. While the geometric data are known in advance the material properties are to be determined first by adequate methods. The parameters of the local friction characteristic, the non-linear contact stiffness as well as the wear coefficient are identified by experiments.

3.1 Identification of Elasticity Modulus and Damping Coefficient

The linear material model, which is used in the following, is not capable of considering non-linear material properties of rubber e.g. the Payne or Mullins effect [4, 18]. The elasticity modulus is assumed to be constant with respect to the excitation strain, the excitation frequency, the deformation history and temperature. However, these assumptions can be justified by the target of the model: the calculation of high-frequency vibrations which are mostly

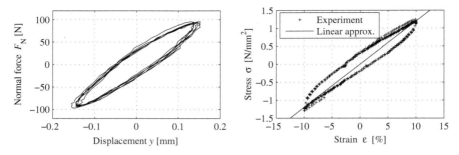

Fig. 11 Left: Force-displacement curve for dynamic excitation at 15 Hz. Right: Quasi-static measured stress-strain curve.

characterised by small vibration amplitudes. Therefore, the constant elasticity modulus can be interpreted as a linearisation in the vicinity of the operating point where the elasticity modulus is approximately constant. Figure 11 (left) shows the force-displacement characteristic of a tread block which is dynamically excited in normal direction with a frequency of $f = 15$ Hz and a strain amplitude of $\epsilon = 3\%$. The elliptic shape points out the linear material behaviour for small deformations [36].

At first, quasi-static experiments from the German Institute of Rubber Technology are employed to determine the elasticity modulus of the elastomer material at a strain of $\epsilon = 10\%$, see Figure 11 (right). A linear regression yields an elasticity modulus of $E = 12$ N/mm^2 for the investigated carbon filled tread compound. In fact, this static value does not hold for highly dynamic processes. The elasticity modulus varies with the deformation velocity and tread block temperature and has to be determined according to the respective state.

In order to set up a structural model for different values of the elasticity modulus the treatment of the stiffness matrix is discussed. The elasticity modulus is at first set to $E = 1$ N/mm^2. The stiffness matrix for an arbitrary elasticity modulus is easily obtained by scalar multiplication of the overall stiffness matrix with the respective value for the elasticity modulus. Within the following, Poisson's ratio ν has been set according to the incompressible rubber material to $\nu = 0.49$. One method to determine the elasticity modulus and the damping coefficient is the analysis of a test with free vibrations of the system. Therefore, the vibrational answer of the system behaviour is analysed with a single-point laser vibrometer. The vibration velocity of the rubber sample is detected with the experimental setup. First measurements with the original rubber tread block geometry provided only unanalysable results due to the small vibration amplitudes which were additionally superposed by higher modes and signal noise. Therefore, another sample geometry has been chosen which allows reasonable measurements.

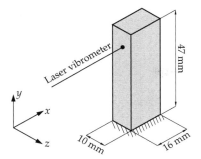 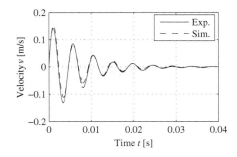

Fig. 12 Left: Experimental setup for the measurement of rubber sample vibrations. Right: Measured and approximated vibration velocity versus time.

The sample has the dimensions 16 × 10 × 47 mm and is glued with the front surface to a rigid base, see Figure 12 (left). The sample is then deflected approximately in the first bending mode shape. The subsequent oscillations show a typical eigenfrequency which can be used for the determination of the elasticity modulus. The sample shape is rather elongate and strongly differs from the tread block shape in order to achieve larger vibration amplitudes and defined mode shapes within the experiment.

The natural frequency of the system is governed by the elasticity modulus, damping, material density, rubber sample geometry and boundary conditions. Except for the elasticity modulus and the damping all other parameters are known. A finite element model with the geometry of the elongate rubber sample has been built up with velocity-dependent damping to approximate the vibrational behaviour and to identify the elasticity modulus. Therefore, the damping matrix **D** is assumed to be proportional to the stiffness matrix **K**

$$\mathbf{D} = \alpha \mathbf{M} + \beta \mathbf{K} \quad \text{with } \alpha = 0. \tag{19}$$

If the damping is at least moderate the undamped natural frequency f_0 is approximately equal to the damped natural frequency $f_{0,D}$. Within the simulation the rubber sample model is deflected in its first bending mode shape similar to the experiment and then fulfils free vibrations. The elasticity modulus E and the damping coefficient β have been found so that the best possible approximation of the vibration frequency and the amplitude decay is achieved. The measured natural frequency of the rubber sample is $f_{0,\text{exp}} 215$ Hz. The best approximation within the model is achieved with an elasticity modulus of $E = 45$ N/mm^2 and a damping coefficient $\beta = 1.5 \times 10^{-4}$ s^{-1}. The agreement with the experiment is acceptable due to the non-linear rubber properties which cannot be covered by the linear material model, see Figure 12 (right). Nevertheless, the identified parameters can be used as guide values. In spite of the different geometry between vibration test sample and the original tread block sample the identified parameters can be transferred to the tread block system because the geometry of both systems is considered by the finite element model within the system matrices.

3.2 Identification of Density

The density of the rubber material has been determined by weighing a defined rubber volume to $\rho = 1.15$ g/cm^3.

3.3 Optimisation of Number of Modes

For the reduction of degrees of freedom which are based on the pure modal or combined static and modal condensation the number of considered modes plays a central role. On the one hand a small number of modes leads to a high reduction rate. One the other hand a higher number of modes leads to a better dynamical approximation of the system behaviour. Unfortunately, there is no general guideline for a reasonable number of modes in advance. Therefore, a parameter variation is conducted: a force step excitation is applied at the first node (bottom left) in horizontal direction at the simulation time $t_0 = 1$ ms. The dynamic system answer is investigated for four different reduction configurations:

- Static condensation,
- Craig–Bampton reduction with 1 mode,
- Craig–Bampton reduction with 5 modes,
- Craig–Bampton reduction with 10 modes.

The simulation results of the reduced systems are compared to the full finite element simulation results which is not subjected to a reduction. Figure 13 depicts the horizontal displacement versus time of the first node, which is the location of the force excitation. Additionally, the results for node 80 in vertical direction (DOF 160) are shown which represents a secondary degree of freedom at the right edge. To obtain the physical displacement of this condensed degree of freedom the backward transformation is necessary. In case of the static condensation (SC) the simulation differs in amplitude and phase from the finite element calculation. As expected, the static contribution is covered correctly.

The consideration of one modal shape function (CB (1 mode)) results in a considerably higher approximation quality. The amplitude and phase behaviour is much better but there are still differences for high-frequency contributions. These differences decrease for five modal shape functions (CB (5 modes)) where only minor deviations from the finite element solutions are observed. In case of ten additional modes (CB (10 modes)) no differences are observed for both exemplarily shown degrees of freedom. In all following simulations the number of modes has been set to 10 which gives a reduction rate of 87%.

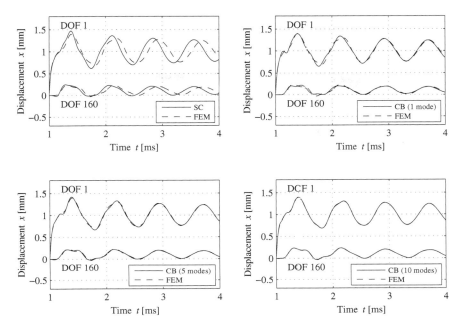

Fig. 13 Influence of the number of modal shape functions on the dynamic behaviour.

3.4 Identification of Local Friction Characteristic

The local friction characteristic describes the stationary friction coefficients in dependence of the parameters relative velocity and nominal contact pressure. As additional important parameter the contact temperature is essential. During the experimental investigations, however, it is not possible to preset the contact temperature. In fact, the contact temperature has to be understood as a resulting inner variable which is influenced by the frictional process. Within the regarded velocity range of the experimental investigations of the local friction characteristic which are shown here (up to 0.3 m/s) a nameable increase of the contact temperature is observed, see Figure 14. Experiments with a small rubber sample on a corundum surface show that the contact temperature increases by about 10 K for this range of sliding velocities. The temperature of the rubber surface is measured with a thermography camera through an elongated hole in the friction surface. The temperature characteristic can be approximated by a power law [27, 29].

Thermographical measurements of a tread block sliding on a germanium surface reveal the temperature distribution in the contact area: the highest temperature occurs at the leading edge where a lip is often observed leading to a high normal contact pressure, see Figure 14 [23]. The experiments shown here for the determination of the local friction characteristic have been

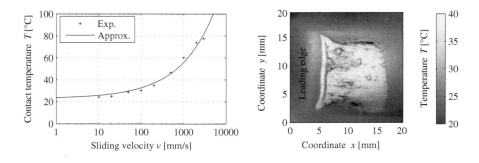

Fig. 14 Left: Measured contact temperature versus sliding velocity. Right: Contact temperature distribution of tread block on a germanium window.

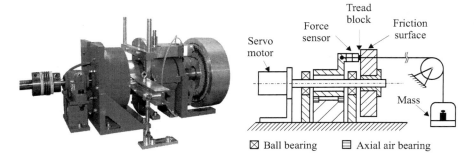

Fig. 15 Tribometer test rig.

conducted on a tribometer test rig at the Institute of Dynamics and Vibration Research at the Leibniz University Hannover [11, 12], see Figure 15. The test rig is a rotational tribometer. The relative velocity between the contact partners is provided by a rotating disc. The disc which is mounted with the friction surface is driven by a servo motor. The rubber sample is fixed onto a sample holder and has one translatory degree of freedom normal to the friction surface in order to apply the normal force by weights.

Figure 16 (left) shows the measured friction characteristic of a small rubber sample with a diameter of 10 mm and a height of 2 mm on a corundum surface grit 400. The friction coefficient decreases with increasing normal pressure which is characteristic for rubber friction. Moreover, the friction coefficient shows a maximum with respect to the relative velocity at about 100 mm/s. The maximum measured friction coefficient is $\mu = 2.48$. Another measurement has been performed on a concrete surface with a tread block with the dimensions $15 \times 15 \times 10$ mm, cf. Figure 16 (right). This characteristic is similar to the measurement on corundum with a decreasing friction

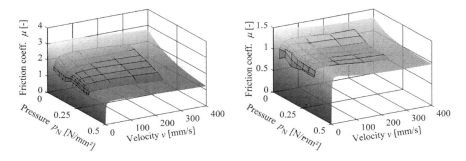

Fig. 16 Measured (grid) and approximated (semi-transparent) friction characteristic of corundum (left) and concrete (right).

Table 1 Identified coefficients for approximation of local friction characteristic

Parameter	Concrete	Corundum 400
k_S [s/mm]	0.8	0.8
$\mu_{0,v}$ [−]	1.22	1.3
$\mu_{\infty,v}$ [−]	1	0.9
γ_v [s/mm]	−0.01	−0.0018
$\mu_{0,p}$ [−]	1.3	2.5
$\mu_{\infty,p}$ [−]	0.95	1
γ_p [mm²/N]	−8	−5

coefficient with increasing normal pressure and a maximum with respect to the relative velocity at about 50 mm/s. However, the friction coefficients are generally lower.

Both measured friction characteristics have been approximated with the function described in Section 2.2. The identified parameters have been found by a fitting algorithm and are given in Table 1.

The corresponding approximated friction characteristics are shown in Figure 16 with a semi-transparent shading.

3.5 Identification of Non-linear Contact Stiffness

The non-linear contact stiffness parameters have been determined experimentally. The tribometer test rig shown in Figure 15 has been modified to measure the static contact stiffness between a tread block and different surfaces in normal direction [10]. Defined weights are used to apply different normal forces. The setup is mounted on air bearings which avoid dry friction to measure very small contact forces accurately. The measurement procedure of the normal force-displacement relation starts with the smallest and ends with

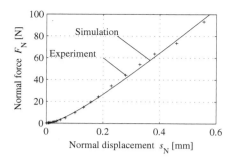 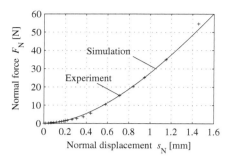

Fig. 17 Measured and approximated normal force-displacement characteristic on corundum (left) and concrete (right).

Table 2 Identified coefficients for the description of normal contact stiffness

Parameter	Concrete	Corundum 400
c_∞ [N/mm^2]	6.25	43
k [mm^{-1}]	0.4	4

the highest normal contact forces. The rubber block is not separated from the rough surface during one complete series of measurements. The normal force-displacement relation $F_N(s_N)$ has been analysed for the corundum and the concrete surface.

The coefficients c_∞ and k which are described in Section 2.3 consider the non-linear contact stiffness. They are found by a fitting procedure to obtain the best approximation of the measurement results. The sum of the normal contact forces in the simulation corresponds to the defined load within the experiment. The comparison of the simulation with the experiment shows a good agreement for the corundum as well as for the concrete surface over the whole measurement range, see Figure 17.

The identified coefficients for the non-linear contact stiffness are listed in Table 2.

3.6 Identification of Wear Coefficients

The identification of the wear coefficient is rather difficult because an adequate wear law has to be found. Here, a simplified wear law is implemented in the model which only has one coefficient to determine. Therefore, an exemplary wear experiment from Gäbel has been chosen to estimate the wear coefficient, see Figure 18 (left) [12]. The measurements have been conducted

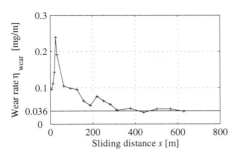

Fig. 18 Left: Measurement of wear rate on concrete surface [11]. Right: Typical tread block shape after wear measurement on corundum.

on the tribometer test rig, cf. Figure 15, with a concrete surface. The relative velocity between tread block and concrete surface is $v_{\text{rel}} = 300$ mm/s and the normal force is $F_N = 45$ N corresponding to a nominal pressure of $p_N = 0.2$ N/mm². After the measurements the tread block reveals a characteristic worn shape, see Figure 18 (right), which is discussed in Section 4.2. The mean friction coefficient for the measurement is $\mu = 0.8$. For these conditions a steady wear rate of $\eta_{\text{wear}} = 0.036$ mg/m is detected for sliding distances $s \geq 300$ m. The corresponding frictional power during the experiment yields

$$P_{\text{fric}} = \mu F_N v_{\text{rel}} \approx 10.8 \text{ Nm/s}. \tag{20}$$

The mass loss per time $\dot{m}$ can also be expressed by the relative velocity v_{rel} and the wear loss per sliding distance and reads

$$\dot{m} = \eta_{\text{wear}} v_{\text{rel}} = k_{\text{wear}} P_{\text{fric}} = k_{\text{wear}} p_N A \mu v_{\text{rel}}. \tag{21}$$

This equation can be solved with regard to the wear coefficient k_{wear}

$$k_{\text{wear}} = \frac{\eta_{\text{wear}}}{p_N A \mu} \approx 0.99 \text{ mg/J}. \tag{22}$$

This experimentally obtained wear rate can now be used as input parameter for the tread block model. However, the discussed wear rate can only be seen as a rough estimation because rubber wear shows a highly non-linear character which is not understood in all its details and not covered by this rather simple wear law.

4 Simulations

Simulations have been performed to study the stationary and instationary behaviour of the tyre tread block. The relevant parameters have been identified in Section 3, so realistic studies of the deformation and normal contact pressure behaviour can be performed. In a further step the influence of wear and the occurrence of high-frequency stick-slip vibrations is discussed.

4.1 Stationary Tread Block Behaviour

In a first study the stationary behaviour is investigated. Therefore, the relative velocity between tread block and surface has been set to $v_0 = 2000$ mm/s. In this velocity range no friction induced vibrations occur due to the nearly constant friction coefficient with respect to the sliding velocity. The compression height H, cf. Figure 10, of the tread block is adjusted to a constant value of $H = 10$ mm which leads to a corresponding nominal pressure of $p_N = 0.25$ N/mm^2. The wear constant has been set to $k_{wear} = 0$ so the tread block keeps its original rectangular geometry. The tread block shows a stationary deformation i.e. the deformed shape does not change with respect to time. Figure 19 depicts the deformation shape of the tread block and the corresponding normal contact pressure. The contact pressure is calculated from the normal contact forces and the corresponding contact area of each point contact element. The tread block experiences a combined compression and shear deformation. The leading edge of the tread block shows a very high normal pressure which leads to very high friction forces which in turn result in a large deformation. The pressure peak originates from the incompressible rubber material property. When the rubber deforms there is only a limited potential to evade leading to high reaction forces in the contact area. The trailing edge has no contact to the surface which is confirmed by the pressure distribution where the local contact pressure is $p_N = 0$ N/mm^2. This behaviour is often observed within experiments where for nearly all load conditions the trailing edge lifts off and still has the original surface, cf. Figure 18 (right).

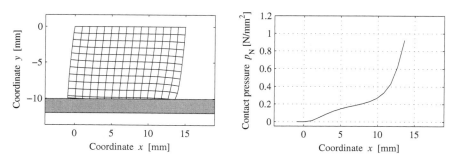

Fig. 19 Deformation of tread block (left) and corresponding contact pressure distribution (right).

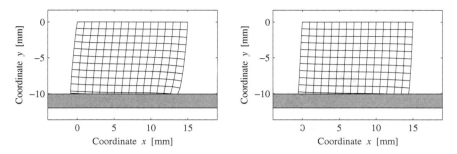

Fig. 20 Deformation of tread block for initial state (left) and after a sliding distance of $s = 10\,\text{m}$ with wear simulation (right).

4.2 Influence of Wear

This section deals with the influence of wear on the friction and deformation behaviour of the tread block. During experiments on rough surfaces it can always be observed that wear changes the tread block geometry. However, tread wear of tyres is a long-term effect. Long simulation times are necessary to clearly observe an influence. Therefore, in the following simulations are performed with a very high wear rate of $k_{\text{wear}} = 15\,\text{mg/J}$ which is about 15 times higher than the identified wear coefficient, cf. Section 3.6. Figure 20 (left) depicts the deformation shapes of the tread block at the beginning of the sliding process where the tread block still has its rectangular shape which corresponds to the simulation without wear. Figure 20 (right) shows the deformed tread block after a sliding distance of $s = 10\,\text{m}$. The compression length H of the tread block has been kept constant. The leading edge deformation reduces because of the high wear at the leading edge whereas the trailing edge is unaffected by wear due to the local separation.

The simulated local wear changes the tread block geometry and, therefore, the resulting pressure distribution. With increasing sliding distance the tread block takes a typical s-shape which leads to a homogenisation of the normal contact pressure. This s-shape behaviour is often observed in experiments, see Figure 18 (right). The tread block shape and the corresponding normal pressure distribution after four different simulated sliding distances are depicted in Figure 21. At the beginning the rectangular tread block shows the above discussed pressure peak at the leading edge. With increasing sliding distance this pressure peak lowers and also the trailing edge contacts the surface. Finally, the contact pressure is nearly constant. Investigations with a thermography camera confirm the homogenisation of the pressure distribution indirectly: at the beginning of the sliding process the temperature distribution shows a peak at the leading edge. With increasing sliding distance a temperature field is built up which corresponds to a constant pressure distribution [23].

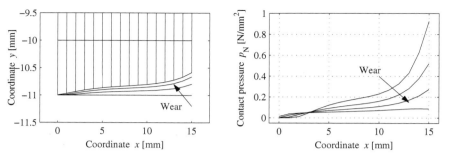

Fig. 21 Effect of wear on tread block shape and contact pressure distribution.

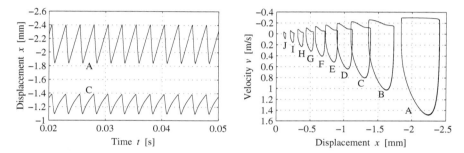

Fig. 22 Displacement of leading edge versus time during stick slip (left) and corresponding phase plot (right).

4.3 Dynamic Tread Block Behaviour

Within the next step the dynamic behaviour of the tread block model is investigated. Here, especially stick-slip vibrations are important which are the origin of tyre squeal. Other dynamic effects resulting e.g. from run-in or snap-out effects are discussed in Section 4.5.

4.3.1 Stick-Slip

Simulations are performed with a sliding velocity of $v_0 = 300$ mm/s and the compression height H is set so that a normal pressure of $p_N = 0.25$ N/mm^2 results. In this range of the sliding velocity the gradient of the friction characteristic with respect to the relative velocity is negative. This leads to an energy input into the system [39]. As a consequence, stick-slip vibrations occur within the simulation with the typical sawtooth shape. Figure 22 (left) shows the horizontal displacements of the nodes A and C, cf. Figure 4. During the sticking phase point C reveals a curved shape in the time signal.

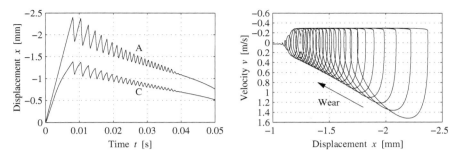

Fig. 23 Displacement of leading edge versus time during stick slip with wear simulation (left) and corresponding phase plot of point A (right).

This indicates a varying velocity during the sticking phase. The stick-slip frequency is about $f_{ss} \approx 400$ Hz. The duration for the simulation is approximately 30 s on a standard desktop PC (Intel Core 2 Duo 2.0 GHz, 3 GB RAM) which points out the numerical efficiency of the model. Figure 22 (right) depicts the phase plot of the simulated stick-slip vibrations for the nodes A to J in horizontal direction. Point A, which is directly in the contact area at the leading edge first sticks to the surface. Therefore, the overall velocity of this point corresponds to the sliding velocity of $v_0 = 300$ mm/s. When the structural restoring forces exceed the friction forces the tread block snaps back with a velocity up to 1.5 m/s until it sticks again, cf. [43]. As already observed in Figure 22 (left) the behaviour of the other points B to J of the leading edge reveal a different behaviour: they show a characteristic overshoot at the beginning of the sticking phase which means that these nodes temporarily provide a higher velocity due to the structural effects of inertia and elasticity which has also been observed in experiments [31].

4.3.2 Stick-Slip and Wear

The effect of wear changes the tread block geometry and the normal contact pressure distribution. Consequently, wear also influences the dynamic behaviour of the tread block. Simulations are conducted with the parameters used in Section 4.1. The load is again defined by the compression height H which remains constant during the simulation. The stick-slip amplitudes as well as the superposed static deformation decrease, cf. Figure 23. The reason is that the resulting normal forces decrease due to the wear effects. The stick-slip frequency increases during the simulation because of the shorter sticking phases. Finally, the oscillations cease which can also be seen in the phase plot.

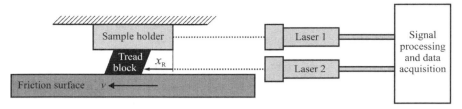

Fig. 24 Experimental setup for stick-slip measurements.

4.4 Comparison with Experiment

Within this section the results of the tread block simulations are compared to experimental results. The measurements have been conducted on the tribometer test rig. The tread block slides on a corundum surface grit 400 which complies with the surfaces which have been used for the determination of the local friction characteristic and the contact stiffness experiments. Pronounced high-frequency stick-slip vibrations occur at a sliding velocity of $v_0 = 300$ mm/s and a normal pressure of $p_N = 0.1$ N/mm^2 which can be acoustically perceived as squealing sound. A single-point laser vibrometer in differential operation mode is used to analyse these vibrations. It measures the velocity and the displacement between two measurement spots. In order to eliminate potential vibrations of test rig components which would superpose the actual tread block vibrations the signals are measured between the tread block and the sample holder. Due to the experimental circumstances the laser spot is focused on the leading edge about 2 mm above the contact surface, see Figure 24.

The displacement as well as the velocity signal show robust vibrations at a frequency of $f_{ss} \approx 1540$ Hz. Figure 25 (left) depicts a cut-out of the unfiltered displacement signal. The corresponding simulation has been performed with the identified parameters of the corundum surface. The relative velocity has been set to $v_0 = 300$ mm/s and the compression has been adjusted so that a mean nominal normal pressure $p_N = 0.1$ N/mm^2 results. The elasticity modulus has been set to $E = 22$ N/mm^2 and the damping coefficient to $\beta = 2.8 \times 10^{-4}$ s^{-1}. The value of the elasticity modulus is between the static value and the identified modulus from the free vibrations experiment. It refers to the frictional heating of the rubber material which is not covered by the model. The comparison of the displacement signal with respect to time is good, see Figure 25 (left). The amplitudes as well as the vibration frequency correspond. Figure 25 (right) shows the phase plot where the velocity of the leading edge v_R is plotted versus the displacement at the leading edge x_R. There are slight differences at the transition from sliding to sticking. Nevertheless, it has been shown by the experiment that the simulation considers the relevant effects with a high numerical efficiency.

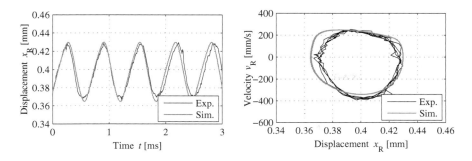

Fig. 25 Comparison of experiment and simulation.

4.5 Rolling Contact

Until now only the pure sliding process of the tread block has been investigated. However, tread blocks perform a superposed motion of translation and rotation under real operating conditions. This effect is considered by the following implementation of the rolling contact.

4.5.1 Rolling Contact Implementation

For the implementation of rolling contact some assumptions are made: The stiff steel belt determines the kinematics of the soft rubber tyre tread. Hence, the single tread block passes a displacement controlled trajectory within the simulation.

Gyroscopic and centrifugal forces acting on a single tread block are neglected because they are small compared to the normal and tangential contact forces. Furthermore, they act evenly on the whole tread block and therefore only marginally influence processes in the contact area and dynamic effects. With these assumptions the rolling motion can be applied to the tread block model without losing numerical efficiency which is one of the basic requirements. The tread block is, therefore, still simulated with a fixed support. However, the contact forces of a tread block performing the rolling motion of the tyre are applied to the model. This approach has the advantage that the above described model with the reduced system matrices as well as the nodal coordinates remain unchanged. Figure 26 depicts the concept of the rolling contact implementation: The fixed system gets as input data the trajectory $y_{Tr}(x_{Tr})$ of the tyre belt for the translatory motion of the tread block. The angle $\beta(x_{Tr})$ describes the gradient of the trajectory and considers the rotatory motion of the tread block. The combined motion of translation and rotation is superposed with the system coordinates $\mathbf{x_{red}}$ of the reduced system. Therefore, also the algorithm for the detection of contact is the same.

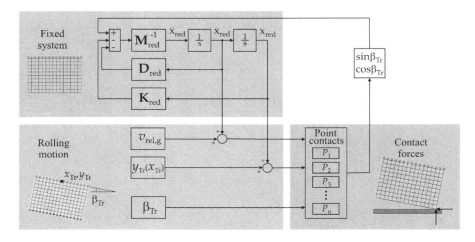

Fig. 26 Scheme of rolling contact implementation.

The point contact elements P_i provide the normal and tangential contact forces which are transformed in the fixed tread block system and act on the respective contact nodes. Additionally, the relative velocity $v_{\text{rel,g}}$ between tyre belt and road enters the simulation and significantly influences the frictional behaviour due to the velocity and pressure-dependent friction characteristic.

This relative velocity has to be defined prior to the simulation and represents the tyre slip velocity. The distance between steel belt trajectory and road surface defines the normal load due to the displacement controlled motion of the tread block on the steel belt trajectory. This distance has to be adjusted according to the respective tyre load conditions. The steel belt deformation is calculated in advance by finite element calculations of a whole tyre.

Here, FE calculations for a truck tyre under stationary rolling conditions are used in the following, see [33]. In order to conform the 3D FE tyre data to the 2D tread block model the centreline in circumferential direction of the outermost belt layer has been extracted. Due to the discretisation of the FE model it is necessary to smooth the belt data in order to obtain a continuous rolling motion. A polynomial approximation leads to good fitting results in the vicinity of the contact patch, cf. Figure 27 (left). The approximation function reads

$$y_{\text{Tr}}(x_{\text{Tr}}) = \sum_{n=0}^{9} a_n \, x_{\text{Tr}}^{2n} \qquad (23)$$

and directly enters the simulation as trajectory coordinates $y_{\text{Tr}}(x_{\text{Tr}})$, see Figure 26. The coefficients of the polynomial function are given in Table 3.

Dynamic Interaction between Tyre Tread Block and Road

Table 3 Coefficients for polynomial approximation of tyre belt data

a_0 [mm]	a_1 [mm^{-1}]	a_2 [mm^{-3}]	a_3 [mm^{-5}]	a_4 [mm^{-7}]
-506.1	0	1.083×10^{-8}	6.308×10^{-13}	-3.328×10^{-17}

a_5 [mm^{-9}]	a_6 [mm^{-11}]	a_7 [mm^{-13}]	a_8 [mm^{-15}]	a_9 [mm^{-17}]
7.444×10^{-22}	-9.2789×10^{-27}	6.663×10^{-32}	-2.578×10^{-37}	4.166×10^{-43}

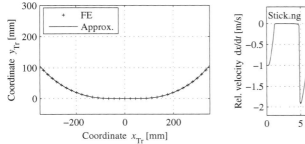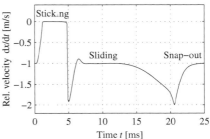

Fig. 27 Left: Polynomial approximation of finite element tyre belt deformation. Right: Relative velocity of leading edge during passage of contact patch.

4.5.2 Rolling Contact Simulations

The following simulations are conducted with the contact parameters and the friction characteristic shown in Figures 16 and 17 and, therefore, represent a corundum surface as contact partner.

The nominal relative velocity between tyre belt and rough surface is $v_{\text{rel,g}} = 1$ m/s which is a typical slip velocity for braking conditions. The distance between the steel belt trajectory and the surface has been adjusted so that a nominal normal pressure of $p_\text{N} = 0.8$ N/mm^2 is applied to the tread block in the middle of the contact patch. This normal load corresponds to truck tyre conditions. During the run-in phase, the tread block contacts the surface with the leading edge which is depicted in Figure 28 with the corresponding contact pressure distribution.

In this state the leading edge experiences a high contact pressure peak whereas the trailing edge lifts off and is not in contact with the rough surface. Here, the tread block is stressed at most. Then a sticking phase follows where the leading edge has no relative velocity to the surface. The tread block shears until the structural restoring forces exceed the friction forces and the tread block comes to the sliding phase. Here, the tread block performs a pure translatory motion where the normal pressure still shows a peak at the leading edge which is lower than in the run-in phase. Finally, the block

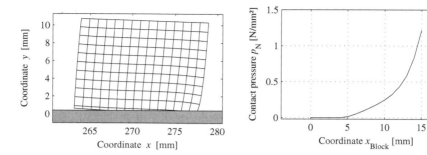

Fig. 28 Deformation of tread block entering the contact patch and corresponding contact pressure distribution.

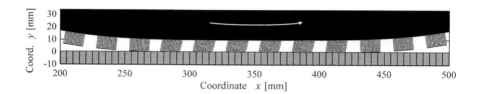

Fig. 29 Chronological view of tread block deformation.

snaps out and returns to its undeformed configuration. For further investigations the relative velocity of the leading edge is shown in Figure 27 (right) where the different phases with sticking, sliding and snap-out can be detected. At the transition from sticking to sliding the tread block is dynamically excited. A chronological view of the deformation behaviour during passage of the contact patch depicts Figure 29.

Figure 30 depicts the resulting forces of the tread block normal ($F_{\text{block},y}$) and tangential ($F_{\text{block},x}$) to the trajectory. They are described in a tread block fixed coordinate system. These forces represent the reaction forces to the tyre belt and can be implemented in global tyre models as excitation forces from the tread blocks without modelling them. Within a second simulation the nominal pressure is reduced to $p_{\text{N}} = 0.2$ N/mm^2 which is a typical value for passenger cars. All other parameters are unchanged. Then another phenomenon is observed: due to the decreasing friction characteristic with respect to the relative velocity friction induced vibrations occur in this parameter range. The displacement of the leading and the trailing edge show the typical saw tooth behaviour, see Figure 30 (right). The whole tread block performs oscillations which are known as stick-slip vibrations. For these conditions there is only a very short sticking phase at the beginning of the contact zone which can be explained by the comparably small nominal normal

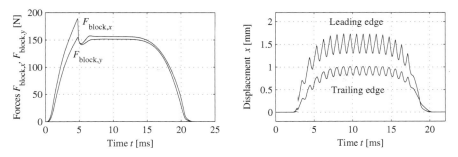

Fig. 30 Left: Resulting reaction forces of the tread block. Right: Stick-slip vibrations of leading and trailing edge.

contact pressure. The stick-slip frequency is $f_{ss} \approx 1200$ Hz which is in the squeal regime. From a safety-related point of view these vibrations are unwanted. They lead to changing contact conditions that influence with a high frequency the contact area of the tread block and the frictional behaviour. This reduces the deceleration during braking and the potential cornering forces.

5 Conclusion

The presented model covers the dynamics of a tyre tread block with a focus on numerical efficiency. The modularly arranged model considers structural effects such as inertia, elasticity and damping, the local friction characteristic, the non-linear contact stiffness and wear effects. The model parameters are mostly gained from experiments. Stationary simulations show the typical deformation behaviour and a characteristic wear shape which is often observed in experiments. Stick-slip vibrations occur in a certain parameter range within the model due to the negative gradient of the friction characteristic with respect to the relative velocity. These high-frequency effects are considered by the model. The agreement between an experimental stick-slip investigation on a corundum surface with simulation results is good. To include the rolling contact the tread block model follows the trajectory of a steel belt. Simulations show four different phases during passage of the contact patch: run-in, sticking, sliding and snap-out. The model simulates tyre tread block squeal effects in a certain contact parameter range which result from friction induced vibrations. The model furthermore delivers the reaction forces to the fixed support which represents the coupling to the tyre belt. The results can be used in global tyre models as excitation forces from the tread blocks without modelling them. This concept allows the realistic and numerically efficient calculation of tyre dynamics.

Acknowledgement

The investigations received financial support from the German Research Foundation DFG within the research group FOR492 "Dynamic contact problems with friction of elastomers".

References

1. Andersson, P.: Modelling interfacial details in tyre/road contact – Adhesion forces and non-linear contact stiffness, PhD Thesis, Chalmers University of Technology, Gothenburg, Sweden (2005)
2. Archard, J.F.: Elastic deformation and the laws of friction. J. Appl. Phys. 24, 981–988 (1953)
3. Bathe, K.J., Gracewski, S.: On non-linear dynamic analysis using substructuring and mode superposition. Computers and Structures 13, 699–707 (1981)
4. Besdo, D., Ihlemann, J.: Directional sensitivity of Mullins effect. In: Constitutive Models for Rubber IV, Balkema, Rotterdam, pp. 229–235 (2005)
5. Bschorr, O., Wolf, A., Mittmann, J.: Theoretische und experimentelle Untersuchungen zur Abstrahlung von Reifenlärm, MBB-Bericht Nr. BB-48381-Ö (1981)
6. Busfield, J., Liang, H., Fukahori, Y., Thomas, A.G.: Modelling the abrasion process in elastomer materials. In: Constitutive Models for Rubber IV, Balkema, Rotterdam, pp. 139–143 (2005)
7. Craig, R., Bampton, M.: Coupling of substructures for dynamic analyses. AIAA Journal 6(4), 1313–1319 (1968)
8. Craig, R.: Coupling of substructures for dynamic analyses – An overview. In: Proceedings AIAA Conference Structures, Structural Dynamics and Materials, Paper 2000-1573, pp. 1–12 (2000)
9. Fleischer, G.: Energetische Methode der Bestimmung des Verschleißes. Schmierungstechnik 4(9), 269–274 (1973)
10. Gäbel, G., Kröger, M.: Non-linear contact stiffness in tyre-road interaction. In: Proceedings 6th European Conference on Noise Control (Euronoise), Paper 118, pp. 1–6 (2006)
11. Gäbel, G., Kröger, M.: Reasons, models and experiments for unsteady friction of vehicle tires. In: VDI-Berichte 2014, pp. 245–259 (2007)
12. Gäbel, G., Moldenhauer, P., Kröger, M.: Local effects between the tyre and the road. ATZautotechnology 4, 48–53 (2008)
13. Greenwood, J.A., Williamson, J.B.P.: Contact of nominally flat surfaces. Proceedings of the Royal Society of London A 295(1442), 300–319 (1966)
14. Grosch, K.A.: The relation between the friction and visco-elastic properties of rubber. Proceedings of the Royal Society of London A 274(1356), 21–39 (1963)
15. Gutzeit, F., Kröger, M., Lindner, M., Popp, K.: Experimental investigations on the dynamical friction behaviour of rubber. In: Proceedings 6th Fall Rubber Colloquium, Hannover, pp. 523–532 (2004)
16. Heinrich, G., Klüppel, M.: Rubber friction and tire traction. In: VDI-Berichte 2014, pp. 341–360 (2007)
17. Hofstetter, K.: Thermo-mechanical simulation of rubber tread blocks during frictional sliding, PhD Thesis, Vienna University of Technology, Austria (2004)

18. Hohl, C.: Anwendung der Finite-Elemente-Methode zur Parameteridentifikation und Bauteilsimulation bei Elastomeren mit Mullins-Effekt, Fortschritt-Berichte VDI, Reihe 18(310), PhD Thesis, University of Hannover, Germany (2007)
19. Hurty, W.C.: Dynamic analysis of structural systems using component modes. AIAA Journal 3(4), 678–685 (1965)
20. Kaliske, M., Näser, B., Müller, R.: Formulation and computation of fracture sensitivity for elastomers. In: Constitutive Models for Rubber IV, Balkema, Rotterdam, pp. 37–43 (2005)
21. Klüppel, M., Heinrich, G.: Rubber friction on self-affine road tracks. Rubber Chem. Technol. 73, 578–606 (2000)
22. Kröger, M., Popp, K., Kendziorra, N.: Experimental and analytical investigation of rubber adhesion. Machine Dynamics Problems 28, 79–89 (2004)
23. Kröger, M., Wangenheim, M., Moldenhauer, P.: Temperatureffekte auf das lokale Reibverhalten von Elastomeren. In: VDI-Berichte, vol. 1912, pp. 271–290 (2005)
24. Kröger, M., Moldenhauer, P., Gäbel, G.: Modular Modelling of Dynamic Systems with Elastomer Contacts. IUTAM Book Series, vol. 3, pp. 277–290. Springer, Dordrecht (2007)
25. Kummer, H.W.: Unified theory of rubber and tire friction, Engineering Research Bulletin B-94. Pennsylvania State University, USA (1966)
26. Le Gal, A.: Investigation and modelling of rubber stationary friction on rough surfaces, PhD Thesis, University of Hannover, Germany (2007)
27. Lindner, M.: Experimentelle und theoretische Untersuchungen zur Gummireibung an Profilklötzen und Dichtungen, Fortschritt-Berichte VDI, Reihe 11, Nr. 311, PhD Thesis, University of Hannover, Germany (2005)
28. Liu, F., Sutcliffe, M.P.F., Graham, W.R.: Modeling of tread block contact mechanics using linear viscoelastic theory. Tire Science and Technology 36(3), 211–226 (2008)
29. Moldenhauer, P., Lindner, M., Kröger, M., Popp, K.: Modelling of hysteresis and adhesion friction of rubber in time domain. In: Constitutive Models for Rubber IV, Balkema, Rotterdam, pp. 515–520 (2005)
30. Moldenhauer, P., Kröger, M.: Vibrations of a tyre tread block under consideration of local wear. In: Proceedings 6th European Solid Mechanics Conference, Budapest (2006)
31. Moldenhauer, P., Kröger, M.: Efficient calculation of tread block vibrations. PAMM 6(1), 317–318 (2006)
32. Moldenhauer, P., Ripka, S., Gäbel, G., Kröger, M.: Tire tread block dynamics: Investigating sliding friction. Tire Technology International, Annual Review 2008, 95–100 (2008)
33. Näser, B., Kaliske, M., Andre, M.: Durability simulations of elastomeric structures. In: Constitutive Models for Rubber IV, Balkema, Rotterdam, pp. 45–50 (2005); and personal communication
34. Nettingsmeier, J.: Frictional contact of elastomer materials on rough rigid surfaces. LNACM, vol. 27, pp. 379–380 (2006)
35. Nickell, R.E.: Nonlinear dynamics by mode superposition. Computer Methods in Applied Mechanics and Engineering 7, 107–129 (1976)
36. Ottl, D.: Schwingungen mechanischer Systeme mit Strukturdämpfung. In: VDI-Forschungsheft, vol. 603. VDI-Verlag, Düsseldorf (1981)

37. Persson, B.N.J.: Rubber friction: Role of the flash temperature. J. Phys., Condens. Matter 18, 7789–7823 (2006)
38. Lorenz, B., Persson, B.N.J.: Interfacial separation between elastic solids with randomly rough surfaces: Comparison of experiment and theory. J. Phys., Condens. Matter 21, 1–6 (2009)
39. Popp, K., Rudolph, M., Kröger, M., Lindner, M.: Mechanisms to generate and avoid friction induced vibrations. In: VDI-Berichte, vol. 1736, pp. 1–15 (2002)
40. Qiu, J., Ying, Z., Williams, F.W.: Exact modal synthesis techniques using residual constraint modes. International Journal for Numerical Methods in Engineering 40, 2475–2492 (1997)
41. Salimbahrami, S.B.: Structure preserving order reduction of large scale second order models, PhD Thesis, Technical University Munich, Germany (2005)
42. Sextro, W.: Dynamical contact problems with friction, 2nd edn. Lecture Notes in Applied Mechanics. Springer, Berlin (2007)
43. Trivisonno, N.M., Beatty, J.R., Miller, R.F.: The origin of tire squeal. Kautschuk und Gummi, Kunststoffe 20(5), 278–288 (1967)
44. Stalnaker, D., Turner, J., Parekh, D., Whittle, B., Norton, R.: Indoor simulation of tire wear: Some case studies. Tire Science and Technology 24(2), 94–118 (1996)
45. Tsihlas, D., Lacroix, T., Clayton, B.: A comparison of two sub-structuring techniques for representing the modal properties of tires. Tire Science and Technology 29(1), 23–43 (2001)
46. Viswanath, N., Bellow, D.G.: Development of an equation for the wear of polymers. Wear 181-183, 42–49 (1995)
47. Wangenheim, M., Kröger, M.: Friction phenomena on microscale in technical contacts with rubber. In: Proceedings 9th Biennial ASME Conference on Engineering Systems, Design and Analysis (ESDA 2008), Paper 59507, pp. 1–6 (2008)
48. Wies, B., Drähne, E., Esser, A.: Produktentwicklung im Zielkonflikt: Einflussparameter zur Optimierung von Bremsverhalten und Fahrstabilität. In: VDI Berichte, vol. 1494, pp. 53–75 (1999)
49. Wriggers, P.: Computational Contact Mechanics. Wiley, Chichester (2002)
50. Zhang, S.: Tribology of Elastomers. Tribology and Interface Engineering Series, vol. 47. Elsevier, Amsterdam (2004)
51. Zheng, D.: Prediction of tire tread wear with FEM steady state rolling contact simulation. Tire Science and Technology 31(3), 189–202 (2003)
52. Ziefle, M.: Numerische Konzepte zur Behandlung inelastischer Effekte beim reibungsbehafteten Rollkontakt, PhD Thesis, Leibniz University Hannover (2007)

Micro Texture Characterization and Prognosis of the Maximum Traction between Grosch Wheel and Asphalt Surfaces under Wet Conditions

Noamen Bouzid and Bodo Heimann

Abstract. A reliable online prognosis of the grip available between tire and road is a feature with high potential for further improvements of the automotive safety. Under wet conditions the grip information can be derived from an online measurement of the influencing parameters, like water film thickness etc. The prognosis systems proposed heretofore are able to describe the hydrodynamic fall of traction at high speeds which depends on the vehicle speed, the water film thickness and the drainage properties of the road surface and of the tire. But, there exits a limitation concerning characterization and influence description of the road micro roughness which determines the friction level at low speeds. The present article deals with rubber friction under wet conditions and at low speeds, which is affected by the micro texture. Experimental investigations are done using the Grosch wheel and several asphalt surface samples. The parameters temperature, speed, wheel load, rubber compound and pavement roughness are considered in order to investigate possible interactions between them in respect to the friction coefficient. A micro texture descriptor with a high correlation to the wet grip at low speed is suggested.

1 Introduction

Long before the automotive engineering it was the road construction that tried to establish the relationship between the road pavement roughness and the traction. Wet conditions play in this context a major role, which can be

---

Noamen Bouzid
Robert Bosch GmbH, Robert-Bosch-Str. 1, 77815 Buehl, Germany
e-mail: noamen.bouzid@de.bosch.com

Bodo Heimann
Institute of Mechatronic Systems, Leibniz Universität Hannover, Appelstr. 11a, D-30167 Hannover, Germany
e-mail: bodo.heimann@imes.uni-hannover.de

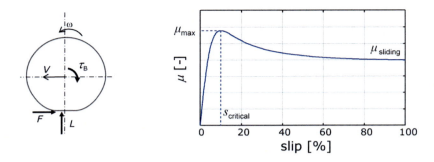

Fig. 1 Braked wheel (left), typical μ-slip characteristics (right).

explained by the fact that under these conditions both extremely high and extremely low friction coefficients occur. In general renewal actions of road pavements are initiated on the basis of measured wet traction. The lack of a reliable description of the relationship between the roughness and the grip necessitates the large-area measurement of the grip by using special vehicles.

Aiming to improve the driving safety the automotive engineering recently has been working on the wet traction prognosis. The terms pavement grip or maximum traction covers as the maximum achievable friction coefficient μ_{max}. Here μ is the ratio of the integral friction force F and the wheel load L

$$\mu = \frac{F}{L}. \quad (1)$$

In case of braking the slip and the braking force, F first increases proportionally with rising braking torque (see Figure 1). Then the μ-slip characteristics take a degressive course until the critical slip/the maximum friction coefficient μ_{max} prevail. When the torque is increased beyond this point, the μ-slip curve passes through an unstable branch. This results in a low sliding friction coefficient.

The maximum traction corresponds to the friction coefficient which occurs during ABS braking. It means that this variable is not instantly measurable. Hence, known analytic approaches [1] which can estimate the tire forces based on a description of the vehicle dynamics are also not applicable. To estimate the maximum traction it is iassumed that it can be derived from the influencing variables. If an unpredictable influence factor with considerable effect is present, as for example the existence of dust or oil on the roadway, the prognosis becomes impossible. In contrast, during or after rain ideal conditions arise for the prognosis. In this case, the influence of the external medium depends only of the water film height. The relationship between μ_{max} and the influencing parameters is described by a data-based model, e.g. by regression models [3, 4].

1.1 Mechanisms of Rubber Friction

The traction between tire and road is established primary through the mechanisms of rubber friction and adhesion and hysteresis [2, 5]. The adhesion friction force is caused by the shearing of the molecular bonds (van der Waals forces) in the contact surface of the friction partners. It is a dissipative stick-slip process at the molecular level [2]. The glue-like connections between the molecular chains of the elastomer and the road pavement are first built locally and then they break. After a brief relaxation phase process repeat cyclically with a frequency given by the sliding speed.

The hysteresis friction arises from the internal damping of the elastomer when excited by the roughness of the road pavement. This friction component is characterized by the cycle of deformation and relaxation of the rubber material through the roughness. The deformed volume does not return completely to the initial state after the relaxation resulting in energy lost and consequently in a friction force.

While the maximum traction μ_{max} on dry road pavements is approximately ranged between 1–1.3, the corresponding range on wet pavements is 0.2–0.9, see [15]. Besides the obvious lower level of traction, a more significant variation is remarkable under wet conditions. This suggests the existence of important effects that must be considered in the first instance for the friction prognosis.

1.2 Maximum Traction under Wet Conditions

Under wet conditions the maximum traction is mainly driven by the hydrodynamic and squeezed film effects. The first effect is due to the increase of the hydrodynamic pressure with increasing speed. The water penetrates gradually in the front area of the contact area between tire and road [2]. This results in a gradual decrease of the traction until the total separation of tire and pavement (aquaplaning). The amount of traction decrease depends on the total duration available which is determined by the driving speed and the duration needed for the elimination of water from the contact area. A high water film, poor drainage capabilities both of tires and roadway and a low wheel load are factors that extend the water removal duration and therefore reduce the traction.

The second main effect that determines the traction under wet conditions is related to the penetration of the last thin water film. Local pressure raises are generated by the micro texture and the sharp edges of the road pavement. The water film is thus interrupted and dry contact areas are formed. The squeezed film effect dominates at low speeds, because the hydro-dynamically related traction decrease occurs only at relatively high speeds ($v > 20$ km/h).

Fig. 2 (a) Grosch wheel, (b) Measuring platform with highlighted Grosch wheel.

Therefore the low speed wet traction is used in the literature often as a descriptor for the road pavement micro roughness [4]. The reversed way, i.e. driving the grip information from a geometrical characterization of the micro texture, is not practicable until now [10].

1.3 Advantage of the Grosch Wheel

The literature in the fields of automotive engineering and road construction consider either measurements at laboratory conditions, where realistic reproduction of the roadway roughness of public roadways is not possible (flat roadway test bench [9], Grosch tester [8]), or street measurements with poor reproducibility. By means of the Reibmobil (measuring platform) friction measurement can be run using the Grosch wheel in the laboratory as well as on real roadways, see Figure 2. The Grosch wheel exhibits stronger synergies to vehicle tires than other specimens because many influence parameters vary in the same way. The dependence of the contact surface on the normal force and the cooling of the contact surface by alternation of contact are two examples of the high similarity. A closer description of the test set-up is available in [11, 12].

An essential advantage of the measuring set-up is the possibility to use real roadway samples in the lab. Since the main focus is set on the analysis of the influence of the micro texture at wet conditions, low speeds are demanded. With this asphalt samples with a length of about 25 cm are sufficient for the measurement. Due to the high reproducibility of the lab measurements, it is possible to investigate the whole process. This is important, because the rubber friction is determined by complicated effects as adhesion and hysteresis which makes interactions between the parameters very potential. The limitation on the variation of the parameters roadway and rubber compound, which is often done in the literature, is insufficient. The present contribution aims a generally valid model of the maximum traction.

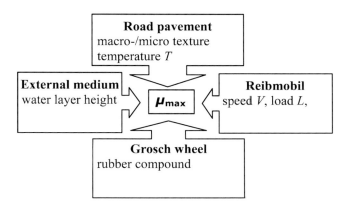

Fig. 3 Parameters influencing the maximum traction between Grosch wheel and road pavement under wet conditions.

2 Experimental Investigation of the Process

The parameters number within the friction process between tire and roadway under wet conditions can be divided into four categories [13]: roadway, vehicle, tire and external medium. Applying this classification to the Grosch wheel results in Figure 3.

The parameters number within the wheel group is reduced in comparison to the tire case because the Grosch wheel has no tread profile or air pressure. Furthermore it has been confirmed, that on account of the low speed the water film height has no influence. Hence six parameters have to be considered.

Table 1 shows the range chosen for the the variation of the different parameters. A high attention was performedpaid to provide test surfaces with diversified texture. Because the samples come from renewed asphalt roadways both worn texture and well-preserved texture is represented, see Figure 4. All three chosen rubber compounds are filled with silica. This filler is used in modern tires, because of the optimum compromise between grip and rolling

Table 1 Parameter variation.

Pavements	Approx. 10 asphalt samples
Rubber compounds	Rubber mixtures N6K, S4K, E6K based on: • Natural rubber (NR) • Styrene-Butadiene (SBR) • EPDM
L	30–150 N
T	5–40°C
V	5–25 mm/s

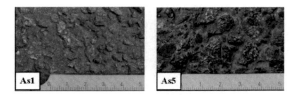

Fig. 4 An example for a sharp (left) and a smooth asphalt sample (right).

resistance. The three produced mixtures exhibit different levels of maximum traction.

The parameters wheel load and temperature are varied in a relatively wide range. The pressure in the contact zone, resulting from the wheel load, is comparable with the pressure in the tire shuffle. The narrow variation range for V results from the restriction to low values described above.

2.1 Reproducibility of the Friction Measurement

Although the measurements are done in the laboratory a good reproducibility is achievable only when special care is taken. Experience has shown that the first measurements should not be considered. The maximum traction then reaches a stationary state. Furthermore applying a short pause of about 10 minutes turned to improve the reproducibility. In addition to that the reproducibility depends on various parameters such as time and the rubber compound.

2.1.1 Impact of Time on the Reproducibility

Table 2 shows the results of a reproducibility study for two rubber compounds. The measurements were carried out in three days. The rubber compound E6K (Table 2a) shows a small dispersion for both the measurements on one day as well as for all measurements. However a deterioration of the reproducibility is generally observed when measurements over several days are involved, see Table 2b. Obviously the rubber compound E6K exhibits optimal properties regarding reproducibility.

These findings can be explained by the change in the mechanical properties of the rubber compound. It is generally known that the mechanical properties of elastomers under dynamic loads vary with the current stress amplitude as well as with the load history. The achievement of a stationary state after few measurements is the result of the formation of a thin rubber layer with altered behaviour. When the measurements are performed without sufficient pause, the properties of rubber in the contact area change continuously. This can be prevented by the 10-minutes pause mentioned above.

Table 2 Impact of time on the maximum traction reproducibility ($L = 70$ N, $V = 5$ mm/s, $T = 18°$C, rough surface: concrete sample, inter medium: water).

(a) Rubber compound E6K			
	1st day	2nd day	3rd day
μ_{max}, five repetitions	0.77	0.79	0.77
	0.78	0.79	0.76
	0.78	0.79	0.74
	0.78	0.76	0.76
	0.77	0.76	0.75
standard deviation for each day	0.006	0.015	0.010
mean value	0.77	0.78	0.75
standard deviation for all three days	0.014		

(b) Rubber compound S4K			
	1st day	2nd day	3rd day
μ_{max}, five repetitions	1.13	1.18	1.08
	1.12	1.17	1.09
	1.08	1.18	1.10
	1.13	1.17	1.13
	1.09	1.17	1.09
standard deviation for each day	0.025	0.005	0.022
mean value	1.11	1.17	1.10
standard deviation for all three days	0.039		

Considering measurements within a longer period the rubber properties converges to different states, resulting in a higher standard deviation of all measurements in comparison to the standard deviation for one day (see Table 2b).

2.1.2 Reproducibility Improvement through Maintaining One Track

The reproducibility of the friction measurement is affected by the small dimension of the Grosch wheel. It exhibits in comparison to the tire a nonuniform force transmission. To illustrate this fact friction measurements on corundum and an asphalt sample are compared. Figure 5 shows the friction coefficient as a function of the distance covered. The measurement is repeated while maintaining the same track. Due to the homogeneity and the fine structure of corundum the small wheel dimension does not have any negative effect. The friction coefficient will be constant regardless of maintaining the same track or not.

The μ profile in Figure 5b can be explained by a fluctuating contact area and thus a variable adhesion friction contribution. In spite of the fluctuation the reproducibility must be comparable to corundum when the track is

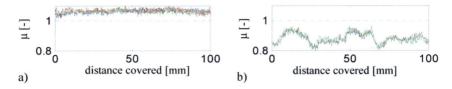

Fig. 5 Friction coefficient as a function of the distance covered, (a) corundum, granularity 240; (b) asphalt sample As2 (rubber compound E6K, $L = 70$ N, $V = 5$ mm/s, $T = 20°$C, inter medium: water).

Table 3 Reproducibility for different tracks (rubber compound N6K, $L = 70$ N, $V = 5$ mm/s, $T = 20°$C, inter medium: water).

	Asphalt sample As5	Asphalt sample As9	Asphalt sample As13
1st track	0.76	0.84	0.78
	0.76	0.83	0.79
2nd track	0.74	0.92	0.79
	0.73	0.93	0.80
3rd track	0.76	0.90	0.83
	0.77	0.91	0.82

maintained during repetitions. Since the average values of the overlapping curves do not scatter significantly.

To demonstrate the influence of different tracks measurements were performed on different parallel tracks. Table 3 shows the average values of the friction coefficient for three different asphalt samples. The reproducibility on one track is for all three surfaces still very good. The maximum deviation between repeated measurements does not exceed 0.01.

However, important deviations exist between the three tracks. The asphalt sample As9 has the largest difference of 0.1. It is assumed that this surface exhibits a non-uniform roughness. Also the two remaining samples confirm the statement that maintaining a track leads to a better reproducibility. The maximum deviations are 0.04 and 0.05 respectively.

2.2 Influence of Wheel Load

By analysing of the normal force dependence the interaction with the influence value "temperature" is to be regarded. By increasing normal force, the dissipated energy increases, leading to a rise of temperature in the contact surface. Because, under practical conditions, the friction coefficient decreases with the temperature, a superposed decreasing effect results. The interdependency with the temperature can be switched off by limitation of sliding

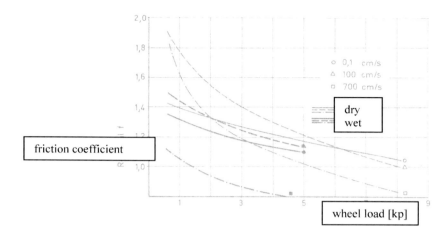

Fig. 6 Friction coefficient depending on normal force for different slide speeds [14].

speed and wheel load. Geyer [14] has determined a fundamental decrease of the friction coefficient with the normal force, during his analyses with square elastic tests (height 1 cm, surface 2 × 2 cm). This trend is valid for all friction surfaces, rubber mixtures and speeds. Indeed, it is strengthened with increasing slide speeds, see Figure 6.

On account of the limited sliding speed, the interdependency of normal force and temperature described above can be eliminated. However, it has to be checked, whether other effects take effect. In this context an adhesion-related effect is to be mentioned. With many surfaces, an increasing normal force causes a disproportionate extension of the contact surface between ground and elastomer. Adhesive friction is proportionally to the contact surface, a decreasing friction coefficient results. Primarily, this effect is provable for idealized surfaces, e.g., for spherical roughness [8]. For dry and plain surfaces an exponent of $-1/3$ occurs with the approach in Equation (2). It has to be shown, whether this takes effect for real asphalt surfaces under the chosen conditions

$$\mu = \mu_0 \cdot \left(\frac{L}{L_0}\right)^n. \qquad (2)$$

The influence of wheel load is analyzed with constant speed ($v = 25$ mm/s) and temperature ($T = 20°C$). As shown in Figure 7, μ_{max} is quasi-irrespective of the wheel load.

The last statement is confirmed by calculating the exponent according to Equation (2), see Table 4. For the rubber mixture on basis of EPDM, a positive exponent is given in general, i.e. μ_{max} increases. Due to the small amount of the exponent, the influence of the wheel load on the capability of adhesion can be neglected. This leads to the assumption, that the contact

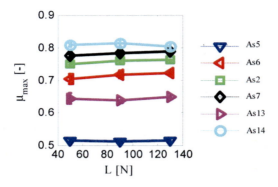

Fig. 7 Influence of wheel load on max exemplarily for the rubber compound E6K.

Table 4 Exponent n according Equation (2), $L_0 = 90$ N, $V = 25$ mm/s, $T = 20°$C.

	E6K	N6K	S4K
As2	0.0164	−0.0131	−0.0126
As5	0.0006	0.0214	0.0722
As6	0.0269	−0.017	0.0032
As7	0.0183	−0.0517	−0.0282
As13	0.0007	−0.0297	0.0357

surface with the real asphalt surfaces is proportional to the normal force. For the remaining analyses the wheel load is set to 70 N.

2.3 Influence of Speed and Temperature

In the 1960s Grosch [5] showed, that measurements of the friction coefficient with different sliding speeds and temperatures, can be converted with the help of the WLF equation in one so-called master curve. The temperature-speed equivalence is valid for a wide temperature range (between 58 and 90°C), different rubber mixtures and friction surfaces, see Figure 8. Besides, the sliding speed was limited to 3 cm/s max., so that no increasing temperature is caused by friction in the contact surface [5]. With these results, Grosch has proved that the main mechanisms of rubber friction, adhesion and hysteresis, are viscoelastic. This means that they are to be led back on the viscoelastic characteristics of the elastomers. If the rubber slides on the street surface, macro and micro roughness cause oscillations on a wide frequency range. Therefore the temperature-speed equivalence is a consequence of the known temperature-frequency equivalence of rubber materials of the dynamic modulus.

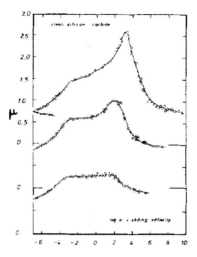

Fig. 8 Master curves of the friction coefficient on a Korund surface. The three curves show the effect of different filler concentrations, see [6].

The analysis of the influence of the temperature or the speed can give an explanation about the impact of the adhesion and the hysteresis. Possible variable influences from both effects would have to be compensated to describe the micro penetration effect correctly.

Temperature and driving speed were varied systematically at a constant normal force. On this occasion, all rubber mixtures and asphalt tests were considered. The WLF relation gives the movement along the logarithmic relative speed axis, if the temperature by the initial value T is being related to the reference temperature T_S,

$$\log a_T = \frac{8.86(T - T_S)}{101.6 + T - T_s}. \quad (3)$$

The reference temperature T_S is a constant of the used rubber. It is ca. 50°C above the glass transition temperature T_G by definition. Table 5 shows T_G and T_S values of the elastomers used. The relative speed results from the product of driving speed and slip

$$v_{\rm rel} = 0.6 \cdot v. \quad (4)$$

The chosen slip of 60% is, in comparison to the critical slip of the tire, very high. The reason is, that the μ-slip characteristic of the Grosch wheel is very low, i.e., it does not have a distinctive maximum. With all considered conditions, a maximum friction coefficient occurs with slip values of 60% [13].

Table 5 Glass transition and correlating reference temperatures of elastomers used.

Elastomer	T_G [°C]	T_S [°C]
EPDM	−45	−5
NR	−35	−15
SBR	−10	40

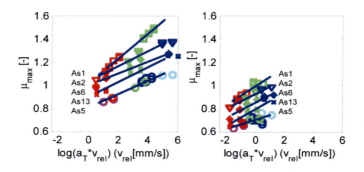

Fig. 9 Master curves for SBR rubber (left) and EPDM rubber (right); referenced to the respective reference temperature T_S.

For all combinations of influence values, a consistent characteristic is shown. The capability of adhesion increases with the variable $v_{rel} \cdot a_T$. In addition, the master curve can be approximated for a specific rubber mixture and asphalt test by a straight line. The master curves of rubber mixtures on basis of SBR and EPDM are shown in Figure 9 as an example.

Due to the restriction on positive temperature values, a relatively small interval of the variables $v_{rel} \cdot a_T$ is given. Looking at the master curves by Grosch in the range between the maximum values of adhesion and hysteresis, the rising course in Figure 9 can be understood. Consequently, with increasing sliding speed or decreasing temperature, max increases due to the hysteresis effect, whereas a reduction of the adhesive contribution takes place.

It is evident that viscoelastic effects are not negligible. On the other hand, these effects seem to influence all asphalt surfaces in similar grade. Figure 9 shows that the characteristics are more or less parallel. Also, the asphalt tests remain in the same order for all rubber mixtures. Based on this statement, a pavement grip index is suggested in Section 3 which is valid regardless of temperature and speed and is to be led back on the micro penetration effect only.

Micro Texture Characterization and Prognosis   213

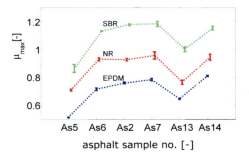

Fig. 10 Maximum traction for all three rubber compounds and divers asphalt surfaces, $T = 18°C$; $V = 25$ mm/s.

2.4 Influence of Rubber Compound

A close correlation between the maximum traction and the glass transition temperature could be observed. Figure 10 shows $\mu_{\max}$ for all three different rubber compounds and various asphalt surfaces.

The rubber compound S4K has the highest maximum traction regardless of the asphalt sample, followed by the compounds N6K and E6K. This corresponds to the order of glass transition temperatures of the corresponding elastomers, see Table 5. To explain this finding the reader is referred to an investigation in [7]. Grosch showed that friction coefficient master curves of various rubbers on glass can be scaled to a common curve when they are referred to the respective reference temperature, see Figure 11b. When an arbitrary temperature is chosen as reference then the curves arrange in the order of the glass transition temperatures, see Figure 11a. This means that

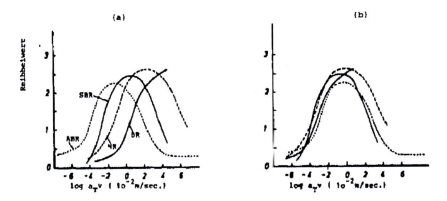

Fig. 11 Friction master curves of several rubbers on glass substrate, (a) referred to 20°C, (b) referred to the respective reference temperature T_S [7].

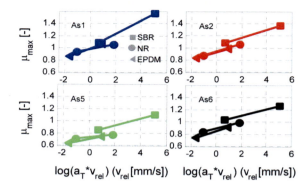

Fig. 12 Friction master curves of all three rubber compounds referred to the respective reference temperature for the asphalt samples As1, As2, As5 and As6.

the correlation between the maximum traction and the glass transition temperature applies for any combination of speed and temperature.

It is important to verify if the last statement still holds when real asphalt surfaces and wet conditions are considered. Each diagram in Figure 12 shows the master curves of the three rubber compounds on a different asphalt sample. Since only a narrow interval of the variable Vrel?aT is covered the curves do not overlap completely. However the curves for EPDM and NR match well. Those both rubbers have comparable glass transition temperatures. The third rubber SBR which exhibits a considerably lower glass transition temperature could be matched if the curves are virtually prolonged. Therefore, the correlation between the maximum traction and the glass transition temperature for any combination of speed and temperature applies also under the present conditions.

3 Pavement Roughness Grip and Grip Index

3.1 The Grip Index

Looking at the μ_{max} level of the various asphalt samples in Figure 10, an obvious similarity of the three rubber compounds can be noticed. If a surface has a relatively high or low maximum traction this holds for all rubbers. Also the temperature and the speed seem to affect the maximum traction in a similar manner regardless of the pavement and the rubber compound, see Figure 12. Both indications suggest that there exist no interactions between the parameter road pavement and the other parameters. In order to prove this supposition μ_{max} was divided by the respective minimum, i.e. $\mu_{max,As5}$. Figure 13 shows the results for all three rubber compounds and two combinations of temperature and speed.

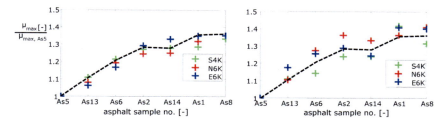

Fig. 13 Grip index, left: $T = T_{\max}/V = V_{\min}$, right: $T = T_{\min}/V = V_{\max}$.

The ratio between the maximum traction of different asphalt samples is nearly constant. The traction measured for any combination of the three parameters rubber compound, speed and temperature varies with the pavement roughness in a unique way. A first application to this finding could be the estimation of $\mu_{\max}$ from the ratio which must be measured in advance (with any settings) and a single measurement at the actual condition. The maximum traction ratio is denoted by grip index or GI:

$$\mathrm{GI} = \frac{\mu_{\max}}{\mu_{\max,\min}}. \tag{5}$$

Since the grip index does not depend on the temperature and the speed, it is supposed to be unrelated to the main mechanisms of the viscoelastic rubber friction adhesion and hysteresis. Also the hydrodynamic effect can be excluded due to the very low speed. Consequently, the grip index can only by affected by the squeezed film effect.

3.2 Characterization of the Pavement Micro Texture

The pavement macro texture can be characterized geometrically. In the recent literature the mean profile depth (MPD) value is often used. According Klempau [4] this descriptor correlates well with the maximum traction at high speeds (hydrodynamic effect). Since no real micro texture descriptor is available, Klempau used a grip characteristic parameter. BFC20: the Brake Friction Coefficient at 20 km/h. This approach – which corresponds to the application proposed in Section 3.1 – involves the disadvantage of using a database including the grip information of the whole road network. Beyond that the data should be updated regularly and must be available for mobile vehicles by wireless technology.

One of the main difficulties of the micro roughness description is the capture of the roughness profile. Laser triangulation sensors (Figure 14) allow its contactless recording. However, these sensors have a limited resolution. The sensor used in the present contribution has a resolution of 0.1 mm. By

Fig. 14 Left: Laser triangulation sensor for friction prognosis [16], right: Laser triangulation sensor used in the measuring platform.

definition the micro texture corresponds to the wavelength range up to 0.5 mm, see [4]. Hence only a limited wavelength range is covered.

Even when new techniques with better resolution are developed in the future, the resolution limitation is required due to the huge amounts of data. This yields a very interesting research topic: whether the maximum traction prognosis is possible in spite of the information lack about the roughness in the range below 0.1 mm.

3.3 Contact Depth Model

The contact depth model after Eichhorn [3] has the advantage of being suitable for online applications due to relatively low computational effort needed. Parameters, such as the contact depth or the contact area are closely related to the pressure in the contact area and can therefore be of interest to describe the squeezed film effect.

The rubber in contact with the road is modelled by a number of springs with the stiffness c and ck. To resolve the contact problem a minimal model with the element i and its neighbours is observed, see Figure 15. The relevant parameters are: the main force F_i, the interlink forces K_{i-1}/K_{i+1} and the partial load L_i.

The partial force results from establishing the force equilibrium for the free cut system:

$$L_i = F_i + K_{i-1} + K_{i+1}. \qquad (6)$$

The internal forces are calculated as follows:

$$F_i = c \cdot d_i, \quad K_{i-1} = ck \cdot (d_i - d_{i+1}) \quad \text{and} \quad K_{i+1} = ck \cdot (d_i - d_{i-1}), \qquad (7)$$

where d_i, d_{i-1} and d_{i+1} denote the respective deformation of the main elements. The two-dimensional contact problem is solved iteratively as follows:

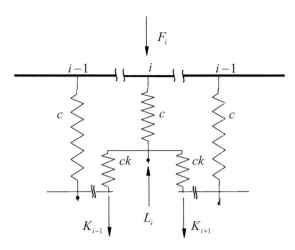

Fig. 15 Contact depth model mechanics [3].

1. the rubber profile which builds initially a straight line is dropped by Δy,
2. the rubber elements having contact with the pavement are identified,
3. the deformations of the elements not in contact are calculated.

The three steps are repeated until the sum of the partial load adds up to the wheel load:

$$\sum L_i = L. \tag{8}$$

The deformation of the elements not in contact are calculated out in sequential steps (e.g. from the left to the right). One step consists of calculating the deformations $d_2 \ldots d_{L-1}$ of the elements between two elements having contact d_1 and d_L according to the following equation:

$$\begin{bmatrix} d_2 \\ d_3 \\ \vdots \\ d_{L-1} \end{bmatrix} = \left(\begin{bmatrix} 1 & -K & 0 & 0 \\ -K & 1 & \ddots & \ddots \\ 0 & \ddots & 1 & -K \\ 0 & \ddots & -K & 1 \end{bmatrix} \right)^{-1} \begin{bmatrix} d_1 \\ 0 \\ \vdots \\ d_L \end{bmatrix}, \quad \text{with} \quad K = \frac{ck}{c + 2ck}. \tag{9}$$

3.4 Correlation between the Grip Index and Contact Depth Model Descriptors

The physical meaning of the pavement micro texture in respect of the maximum traction under wet conditions is the generation of local pressure peaks. Through the pressure peaks the liquid film is interrupted and local dry contact points are built. A noticeable correlation was found between the contact

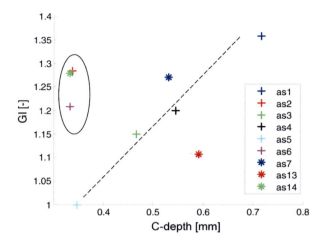

Fig. 16 Correlation between the contact depth and the grip index.

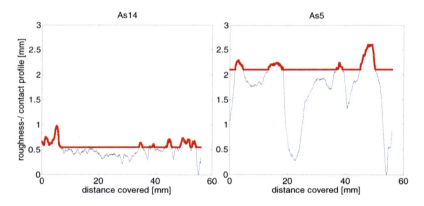

Fig. 17 Roughness and contact profiles for a fine and a coarse asphalt surface respectively.

depth C_{depth} and GI for the majority of the asphalt samples available, see Figure 16. This can be explained by the fact that pavement sharp edges favour the penetration of the rubber.

Considering the three asphalt samples which do not show this correlation (As2, As6 and As14), it can be noticed that those are composed of fine aggregates. Consequently it must be taken into account that such pavements exhibit a high number of contact points and therefore small contact depths. Figure 17 shows the texture and contact profiles of the asphalt surface As14 in comparison with the smooth sample (on the micro scale) As5. As14 exhibits more than twice as many contact points as As5 due to its fine aggregates.

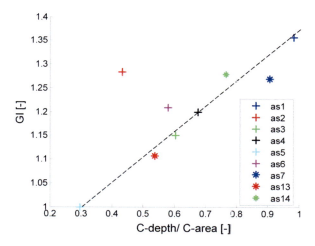

Fig. 18 Squeezed film effect characterized by means of the new micro texture descriptor.

Asphalt pavements with fine aggregates are often deployed in slow driven roadways. The reasons for this are: Firstly the bad drainage properties are not fatal (no hydrodynamic effect at low speeds) and secondly the sharpness of pavement is assured over a longer period. By contrast road surfaces with coarse aggregates tend to be polished. The contact area rises leading the contact pressure to decrease and the squeezed film effect to be weak. This fact is taken into account by dividing the contact depth by the contact area. The result is shown in Figure 18. A high correlation exists between the new descriptor and the grip index. The correlation coefficient is 76% in spite of a clear outlier (As2). Without this asphalt sample the corre- lation would be very high (95%).

4 Conclusion

The existence of a pavement grip index was proved, which depends neither on the viscoelastic friction mechanisms adhesion and hysteresis, nor on the hydrodynamic effect. Thisis index describes the influence of the micro texture on the maximum traction underunder wet conditions.

Thanks to the Grosch wheel investigations, a micro texture descriptor was developed, which describes the squeezed film effect and should allow the prognosis of the maximum traction for the tire in the future. The contact depth model by Eichhorn was applied to estimate the contact depth and the contact surface. The pavement grip index correlates well with the quotient of contact depth and contact surface. This new texture descriptor contains sufficient information about a part of the micro texture and macro texture, to describe the squeezed film effect. Consequently the automotive prognosis of the maximum traction is possible without capturing the roadway roughness in a wavelength range less than 0.1 mm.

References

1. Küçükay, F., Boßdorf-Zimmer, B., Frömmig, L., Henze, R.: Real-time-processing road friction and driving condition estimation. In: 15. Aachener Kolloquium Fahrzeug- und Motorentechnik, Echtzeitfähige Reibwert und Fahrzustandsschätzung (2006)
2. Moore, D.F.: The Friction of Pneumatic Tyres. Elsevier, Amsterdam (1975)
3. Eichhorn, U.: Reibwert zwischen Reifen und Fahrbahn Einflussgrößen und Erkennung, Fortschritt-Berichte VDI, Reihe 12, Nr. 222, VDI-Verlag, Düsseldorf, PhD Thesis, TU Darmstadt (1994)
4. Klempau, F.: Untersuchungen zum Aufbau eines Reibwertvorhersagesystems im fahrenden Fahrzeug, Fortschritt-Berichte VDI, Reihe 12, Nr. 576, VDI-Verlag, Düsseldorf, PhD thesis, TU Darmstadt (2004)
5. Grosch, K.A.: The relation between the friction and visco-elastic properties of rubber. Proceedings of the Royal Society, A 274, 21–29 (1963)
6. Grosch, K.A.: Some factors influencing the traction of radial ply tires. Rubber Chem. Technol. 57 (1984)
7. Grosch, K.A.: Visko-elastische Eigenschaften von Gummimischungen und deren Einfluss auf das Verhalten von Reifen. Kautschuk, Gummi, Kunststoffe 42, 745–751 (1989)
8. Grosch, K.A.: Laborbestimmung der Abrieb- und Rutschfestigkeit von Laufflächenmischungen – Teil I: Rutschfestigkeit. Kautschuk, Gummi, Kunststoffe 6, 432–441 (1996)
9. Roth, J.: Untersuchungen zur Kraftübertragung zwischen Pkw-Reifen und Fahrbahn unter Berücksichtigung der Kraftschlusserkennung im rotierenden Rad, Fortschritt-Berichte VDI, Reihe 12, Nr. 195, VDI-Verlag, Düsseldorf, PhD Thesis (1993)
10. Huschek, S., Merzoug, P.: Zusammenhang zwischen Rauheit und Griffigkeit, Technical Report Heft 735, Institut für Straßen- und Schienenverkehr (1995)
11. Bouzid, N., Heimann, B., Trabelsi, A.: Empirische Modellierung des Reibwertes für das Groschrad-Fahrbahn-System. In: 10. Internationaler Kongress Reifen-Fahrwerk-Fahrbahn, VDI-Berichte 1912, Hannover, pp. 291–307 (2005)
12. Bouzid, N., Heimann, B., Blume, H.: Friction coefficient prognosis for the grosch-wheel. In: Wriggers, P., Nackenhorst, U. (eds.) Analysis and Simulation of Contact Problems. LNACM, vol. 27, pp. 291–307. Springer, Heidelberg (2006)
13. Bouzid, N., Heimann, B.: Friction of rubber wheels on wet asphalt surfaces. In: Proceedings of the 9th Biennial ASME Conference on Engineering Systems Design and Analysis ESDA 2008, Haifa (2008)
14. Geyer, W.: Beitrag zur Gummireibung auf trockenen und insbesondere nassen Oberflächen, PhD thesis, T.H. München (1971)
15. Société de Technologie Michelin, Der Reifen, Haftung – was Auto und Straße verbindet (2005)
16. Trabelsi, A.: Automotive Reibwertprognose zwischen Reifen und Fahrbahn, Fortschritt-Berichte VDI, Reihe 12, Nr. 608, VDI-Verlag, Düsseldorf, PhD Thesis, Universität Hannover (2005)

Experimental and Theoretical Investigations on the Dynamic Contact Behavior of Rolling Rubber Wheels

F. Gutzeit and M. Kröger

Abstract. The simulation of the dynamic rolling contact opens numerous benefits by saving development time and costs. In this chapter, fundamental algorithms for the numerical efficient simulation of rolling rubber wheels are shown. The exposed model is validated by steady and unsteady experimental investigations. For this purpose, a moving test rig has been developed and built up. Outcomes of steady rolling experiments are shown firstly. The resulting friction characteristics reveal the behavior of different elastomers in contact with a glass surface. Furthermore, the dynamic dependency of the tangential force on the start and the target level of a preceding slippage-step is investigated. Based on the dynamic behavior of the rolling contact observed in the measurements, a mechanical model for simulation is presented. The 2D model includes a structure model of the wheel body, as well as a contact formulation in normal and tangential direction reproducing sticking and sliding behavior, respectively. The results of the simulation are opposed to the measurements done with the moving test rig.

1 Introduction

Rolling contacts appear in various fields of technical applications, see Figure 1. Similar to the rolling contact of tyres expanded contact areas also occur between press cylinders or between band conveyors and drive drums.

F. Gutzeit
Institute of Dynamics and Vibration Research, Leibniz University Hannover, Appelstr. 11, 30167 Hannover, Germany
e-mail: `florian@gutze.it`

M. Kröger
Institute of Elements, Design and Manufacturing, Technical University Bergakademie Freiberg, Agricolastr. 1, 09596 Freiberg/Sachsen, Germany
e-mail: `kroeger@imkf.tu-freiberg.de`

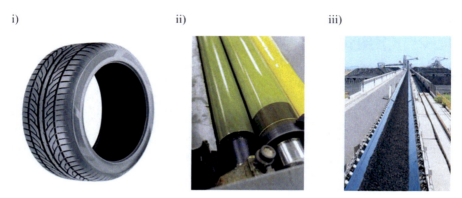

Fig. 1 Fields of application for the unsteady rolling contact model. (i) Contact between tire and road. (ii) Drum contact in printing machines. (iii) Contact between drive drum and band conveyor.

Within this expanded contact area a sticking and a sliding zone establish during the actuated rolling process. An unsteady rolling contact model describes the processes within the contact zone after quick changes of the slippage or of the normal force of the wheel. Thus, the results of this chapter can not only be transferred to the field of tire technologies, but also to many other technical applications. The presented model is realized for a solid rubber wheel (*so called Grosch wheel*), which has a diameter of 80 mm.

Forces and moments transferred by the rolling contact area between tire and road serve for supporting, guiding, actuating and braking purposes. If the driver of the vehicle performs a brake manoeuvre or a rapid change of the steering angle, these contact forces do not follow the driver's demand instantaneously, but with a certain time delay. An efficient modeling of the rolling contact provides many options to gain a deeper understanding of the dynamic processes between tire and road surface as well as in other technical applications.

2 Measurements

To investigate the unsteady behavior of the rolling contact of a so called *Grosch wheel*, a mobile test rig was developed and built up at the *Institute of Dynamics and Vibration Research* of the Leibniz University Hannover. The mobile test rig measures the contact forces of the slipping Grosch wheel within a laboratory environment as well as on the asphalt surface of a public road. The construction and the functionality of the mobile test rig are shown in Section 2.1. Preceding steady friction investigations with the Grosch wheel have proven that the resulting characteristics show a strong similarity to typical tire diagrams, cf. [17]. Nevertheless, a direct transferability is not ensured, because the local contact behavior of a profiled tire clearly differs from that of a Grosch wheel. But the basic processes between rolling wheel and road surface, as the influence of the rolling kinematics or the local friction, are represented sufficiently by the Grosch wheel. Thus it is suitable to systematically investigate the dynamic rolling contact of rubber within a laboratory environment.

2.1 Moving Test Rig

For the experimental investigation of the tire-road contact a mobile test rig was built up [1, 10]. Its concept is based on a predecessor [2, 3], but in addition it was developed to particularly serve for unsteady measurements. A further development objective was to have enough space around the Grosch wheel in order to observe the rolling contact using optical methods (HiSpeed, Infrared camera). During a measurement the test rig moves along with a constant velocity, while the angular velocity and the slippage of the wheel, respectively, follows an given time series. The time series of the normal load of the wheel is realized by a controller as well and can therefore be influenced arbitrarily, too. All forces and moments generated in the contact zone are recorded, the temperatures of the wheel and the road, and the texture of the contacted surface is measured as well. The chassis consists of a torsional stiff aluminium framework, it carries the central process computer and the 60 V energy supply, made up of five battery-blocks. The on-board voltage was increased compared to the predecessor to especially provide the Grosch wheel actuators with more power. The test rig is driven by two hub drives supplied by wheel chair technologies. They were delivered with an analogous joystick, which allows to maneuver the test rig very comfortably and to move forward using selectable speed levels. The actuating elements needed for the measurements and the sensor technology with its processing units are combined in the measurement capsula, which is located in the back of the moving test rig. The measurement capsula can be dismantled by hand so that its only connection to the chassis is by wire. This way it is possible to operate the capsula within other test rigs.

A solid rubber wheel with a width of 18 mm, an inner diameter of 35 mm and an outer diameter of 80 mm serves as a specimen in the measurements. In the tire industry the wheel is used in an apparatus called *Grosch Abrader* for wear testing, see [9]. The composition of the rubber mixture can be chosen arbitrarily, further the body and the mantle (i.e. the outer 5 mm of the wheel structure) can consist of different materials. The wheel is mounted on an adapter, see Figure 2(i), so that it can be quickly fixed to the hub of the measurement capsula by means of an hydraulic clamping bushing.

Unlike the steady measurements, where the influence of wheel load and slippage on the tangential force is described with averaged characteristics, in the analysis of the unsteady measurements the interpretation of time series takes the center role. Since this measurements are more sensitive to disturbances

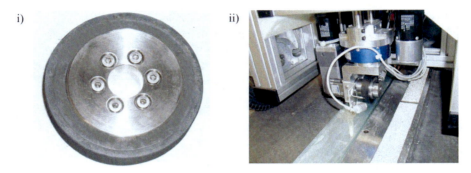

Fig. 2 (i) Grosch wheel, mounted on a wheel adapter. (ii) View on the mounted wheel beneath the measurement-capsula during an experiment on the carrier track.

than the steady tests, the experiments regarding the unsteady rolling contact are carried out in the isolated atmosphere of a measurement hall.

A 2 m long carrier track is installed on the floor of the hall. The carrier track holds selectable surfaces, like for instance glass or corundum, as contact partner for the rubber wheel, see Figure 2(ii).

2.2 Steady Measurements

After the buildup of the test rig, steady measurements varying the normal force F_N and the slippage S were carried out in the first instance. The global slippage is determined by the angular velocity Ω of the Grosch wheel and the velocity v_C of the moving test rig

$$S = \frac{v_C - \Omega r_M}{v_C}, \qquad (1)$$

with the outer radius r_M of the Grosch wheel. Further slippage definitions are mentioned in [25]. The investigations in this chapter were done with different rubber materials, glass was taken as contact partner. Previously experiments with other trackways like corundum, steel and safety walk were accomplished. In view of the interpretation of the time series in the framework of the unsteady measurements, the experiments on glass showed the slightest fluctuations of the tangential force $F_T(t)$. In the measurements the tangential force F_T is recorded in order to calculate the global coefficient of friction

$$\mu^* = \frac{|F_T|}{F_N}. \qquad (2)$$

The global coefficient of friction is marked with asterisk to distinguish it from local sliding friction coefficient μ. It describes both, the sliding friction and also the elastic forces within sticking zone of the contact area. During

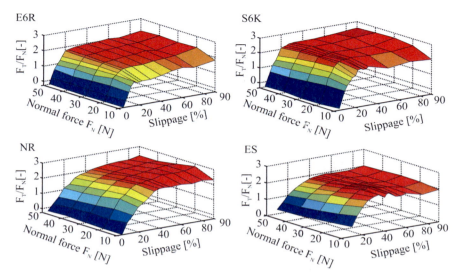

Fig. 3 Rolling friction characteristics for the filled materials E6R, S6K and for the unfilled elastomers NR and ES. All measurements were carried out on a glass surface, at a vehicle velocity $v_C = 40$ mm/s (skew angle $\alpha = 0°$).

the measurements the test rig moves along with the constant velocity $v_C = 40$ mm/s. For every operating point the input parameters slippage S and normal force F_N are kept constantly for the length of 1 m by means of closed loop controls. Subsequently an average is calculated for every section, representing the function value of the operating point $F_T(S, F_N)$. Exclusively braking slippage $S > 0$ was investigated. Figure 3 shows four characteristics resulting from steady measurements with different rubber materials. In all characteristics the slippage is varied in the range of $S = 10, 20, \ldots, 90\%$, with a wheel load of $F_N = 10, 20, \ldots, 50$ N. In all experiments the skew angle amounts to $\alpha = 0°$.

Among the tested materials there are two unfilled rubbers, one natural rubber (NR) and a mixture called ES, consisting of the two synthetic rubbers EPDM (Ethylen-Propylen-Dien-rubber, mass fraction 40%) and SBR (Styrol-Butadien-rubber, mass fraction 60%). For a comparison additionally two filled rubbers were chosen, which are enriched with filler mediums common in the tire industry. The mixture E6R is based on an EPDM polymer, which was filled with 60% phr carbon black. The material S6K consists of an SBR, which was filled by 60% phr silica.

In the measurements, the dependency of the proportion F_T/F_N on the slippage S and the wheel load F_N was investigated. In contrast to the global friction coefficient μ^*, the force proportion F_T/F_N represents a change of sign of the tangential force F_T. All the characteristics show a strong dependency of the proportion F_T/F_N on the slippage S, which complies with a saturation curve. The dependency on the normal force F_N is less distinctive. The maxima of the characteristics are located at $\max(F_T/F_N) \approx 2\text{--}3$. The dependency on the slippage can be divided into a quasi-linear part in the range of lower slippage $S < 20\text{--}40\%$ and a saturation part above. The principle characteristic $\mu^*(S)$ can be explained with basic steady tire models, cf. [5, 15, 19, 23, 28].

The demonstrative difference between the characteristics of the filled and the unfilled materials is the initial slope. In the range of lower slippage the filled materials show an viewable stronger increase than the unfilled rubbers. Thus the saturation limit for E6R and S6K is already reached at a slippage level of $S \approx 20\%$, while for the unfilled rubbers NR and ES the saturation value is achieved for a slippage level of $S \approx 40\%$. The reason for the stiffening of the elastomers is the addition of filler materials, the structure damping is significantly increased as well. This stiffening leads to a stronger increase in the lower part of the $\mu^*(S)$-characteristic, which is dominated by the elastic shear behavior of the wheel.

In the unsteady measurements the dynamic dependency of the tangential force $F_T(t)$ on steps of the slippage $S(t)$ is observed. For this purpose a material with a preferably large range of ascending tangential force $F_T(S)$ is beneficial. Thus steps within a wide range of ΔS can be investigated, whose influence on the time characteristic $F_T(t)$ can still be measured. Since the unfilled rubber materials show a larger range of ascending slippage values, these rubbers are more suitable for the unsteady measurements. Because of the slightly higher values F_T/F_N natural rubber NR is chosen to be the material used for the unsteady measurements.

2.3 Unsteady Measurements

In this section firstly the calculation of a time constant for the characterization of the dynamic behavior of the rolling contact is explained. By means of the identified time constants insights into the dynamics of rolling friction can be gained. As a result of the unsteady measurements the calculated time constants are finally presented for various slippage steps.

2.3.1 Assessment of Time Constant τ

The tangential force $F_T(t)$ does not follow instantaneously the excitation by a slippage step, but with a certain time delay. For the experimental studies of the rolling contact this dynamic behavior is approximated by an exponential saturation characteristic. For the time response of the tangential force immediately after the slippage step at the point in time $t = T_{\text{step}}$ the characteristic

$$F_T^{\text{exp}}(t) = F_T^0 + (F_T^\infty - F_T^0)\left(1 - \exp\left[\frac{-(t - T_{\text{step}})}{\tau}\right]\right) \qquad (3)$$

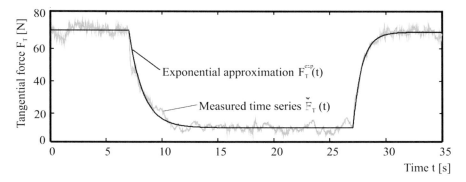

Fig. 4 Comparison between a measured tangential force $\check{F}_T(t)$ and a fitted exponential curve $F_T^{\text{exp}}(t)$ after a slippage step $S = 30 \rightarrow 10 \rightarrow 30\%$ (wheel load $F_N = 40$ N).

is applied. The parameter F_T^0 can descriptively interpreted as the function value before the step $t = T_{\text{step}}$. The asymptote of the saturation characteristic is represented by the parameter F_T^∞. The third parameter τ is associated with the initial slope of the exponential curve. From the point of view of control engineering, where the exponential saturation characteristic is known as the step response of a PT_1-system, τ is denoted as *time constant*. The exponential characteristic in Equation (3) is fitted to the measured tangential force time series $\check{F}_T(t)$ (comprising of m measuring points) by varying this three parameters. As a optimization criterion the *Least-Squares method*

$$\min_{F_T^0, F_T^\infty, \tau} \sum_{j=1}^{m} (F_T^{\text{exp}}(t_j) - \check{F}_T(t_j))^2 \qquad (4)$$

is applied, cf. [13, 16]. The needed start values for the force referenced parameters $^0F_T^0$ and $^0F_T^\infty$ can be taken from the steady measurements. For the time constant the initial value $^0\tau = 1$ s is estimated. With the given start values an algorithm seeks the optimal combination of the three parameters, for which, according to the *Least-Squares method*, the sum of the error squares shows the minimum value. The results lead to a good approximation of the measured date, see Figure 4.

2.3.2 Unsteady Experimental Results

The dependency of the system dynamics on the start and the target level of the slippage step ist regarded in terms of a parameter study. Thereby, the start value of the slippage step is denote as S_{start}, while the target level is characterized by S_{target}. Like in the previous section the vehicle velocity in this experiments is $v_C = 40$ mm/s and the normal force of the wheel

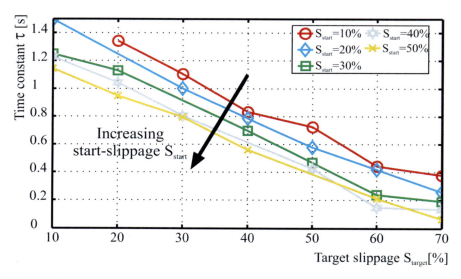

Fig. 5 Time constants of the tangential force $\check{F}_T(t)$ after a slippage step $S_{\text{start}} \to S_{\text{target}}$. The target level S_{target} of the slippage step is represented on the abscissa, the start value S_{start} is regarded as a parameter.

amounts to $F_N = 40$ N, all tests were carried out with the contact pairing natural rubber/glass.

For the start and the target levels of the slippage step the measuring points $S_{\text{start}} = 10, 20, \ldots, 70\%$ were chosen. Above the slippage value $S = 70\%$ no repeatable measurements were possible due to strong damages of the wheel body. Like in the steady measurements in Section 2.2 exclusively longitudinal slippage ($\alpha = 0°$) was regarded. For every measuring point up to 10 experiments were carried out. The resulting time constant τ was calculated by averaging the individual results. In this examination the dependency of the time constants τ on the start and the target level S_{start} and S_{target}, respectively, is regarded. For the separate analysis of the two dependencies the calculated time constants are plotted in two different diagrams, see Figures 5 and 6.

In Figure 5 the time constants τ are plotted versus the target level S_{target} of the slippage step. The start value S_{start} is regarded as a parameter. The diagram shows a set of curves $S_{\text{start}} = 10, 20, \ldots, 50\%$, each curve is composed of up- and down-steps starting from the same slippage level S_{start}. The smallest step height amounts to $\Delta S = 10\%$. For the "not-step" $S_{\text{start}} = S_{\text{target}}$ no time constant can be specified. For all curves an approximate linear, significantly decreasing dependency for increasing values of the target slippage S_{target} is observed. No qualitative difference between up- and down-steps can be noticed, both parts of the curves lie on the same virtual line.

Furthermore a weaker dependency of the time constant τ on the start value S_{start} of the slippage step is identified. This tendency is marked in Figure 5

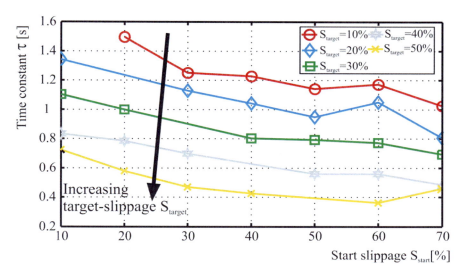

Fig. 6 Time constant of the tangential force $\check{F}_T(t)$ after a slippage step $S_{\text{start}} \to S_{\text{target}}$. The start level S_{start} of the slippage step is represented on the abscissa, the target value S_{target} is regarded as a parameter.

by an arrow. Aside from a few outliers the curves run nearly parallel. The difference between the curves with the start slippage values $S_{\text{start}} = 10\%$ and $S_{\text{start}} = 50\%$ amounts to approximately $\Delta\tau \approx 0.3$ s. The average slope of the characteristics is $m_{\tau,\,target} \approx -2$ s. Slippage steps down to a level of $S_{\text{target}} = 10\%$ result in a time constant between $\tau_{50,\,10} \approx 1.15$ s and $\tau_{20,\,10} \approx 1.45$ s. The lowest time constant within the observed measurement range originates from the step up to the target level $S_{\text{target}} = 70\%$, the values amount to $\tau_{50,\,70} \approx 0.05$ s and $\tau_{10,\,70} \approx 0.4$ s.

To get a more clear image of the dependency on the start value of the slippage step, Figure 6 shows of time constant τ plotted versus S_{start}, here the target value S_{target} serves as a parameter.

The already mentioned weak decreasing dependency of the time constant τ on the start level S_{start} can be obviously seen in Figure 6. Within the observed measurement range this tendency can be regarded as linear. The accordant slope value is $m_{\tau,\,start} \approx -0.8$ s. This is clearly below the slope value $m_{\tau,\,target} \approx -2$ s describing the dependency on the target value S_{target}, which can be seen in Figure 5.

3 Rolling Contact Model

In this section a mechanical model will be shown, which represents the global dynamics of the rolling contact observed in the measurements, see Sections 2.2 and 2.3. Furthermore it gives insights into the local friction behavior, displacements, stresses and relative velocities within the contact zone. Initially the structure modeling based on the finite-element-method is explained. The arising high number of degrees of freedom (DOFs) is strongly reduced by means of the Craig–Bampton condensation. In a static precalculation eventually the initial conditions for the time-step-integration are determined. Preparatory work has been carried out in [11, 12].

During the time-step-integration, the mantle nodes actually next to the contact area are assorted. For each of this nodes a request is done, wether it is touching the track and, in case of contact, if sticking or sliding behavior is existent. As a result of the simulation all required state variables are available on the mantle nodes.

3.1 Efficient Structure Modeling

In this section the input and system matrices needed for the simulation of the behavior of the wheel structure are derived. The equations of motion for the approximation of the dynamic structure behavior are set up by means of the finite-element method. A mixed static and modal reduction algorithm tracing back to [4] leads to a strong reduction of the number of DOFs and consequently to a shortening of the calculation time. Before the start of the simulation the global wheel load as well as the corresponding distribution of the contacting nodes are calculated in a static precalculation. The results of this contact calculation serve as initial conditions of the dynamic calculation.

3.1.1 Kinematics and Dynamics

The overall motion $\delta(t)$ of a rolling, deformed wheel is divided into a given rigid body motion $\lambda(t)$ and small relative displacements $\Delta\delta(t)$ of the nodes. In the numeric simulation, the small relative displacements can be approximated by a linear FE model. The sum of the rigid body motion $\lambda(t)$ and the relative displacements $\Delta\delta(t)$ results in the final configuration

$$\delta(t) = \lambda(t) + \Delta\delta(t), \tag{5}$$

which describes the motion of material points due to external forces. The superposition of the configurations as well as the motion of the center C of the wheel is shown in Figure 7.

The rigid body motion $\lambda(t)$ is determined by the global quantities horizontal hub velocity $^I v_C(t)$ and angular velocity $\Omega(t)$ of the wheel, which are input variables of the simulation model. At the start $t = 0$ the wheel stands still $\Omega(t = 0) = 0$, $^I v_C(t = 0) = 0$. In the vertical direction the rigid body motion of the wheel is determined by the hub displacement $^I w_C(t)$. This quantity depends on the normal force $f_N(t)$ and is computed by a feedback within the model. For the FE-modeling of the structure the rigid body motion $\lambda(t)$ is characterized as reference configuration, which serves as a basis for the

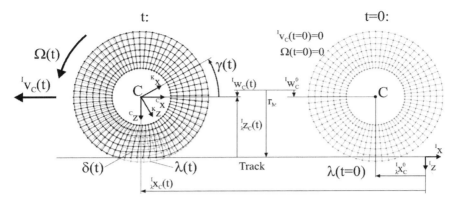

Fig. 7 Wheel in the initial position $t = 0$ (right) and rolling, at the time t (left). The reference configuration $\lambda(t)$ represents the pure rigid body motion of the wheel. The end configuration $\delta(t)$ results by superposing the relative displacements $\Delta\delta(t)$.

dynamic displacements $\Delta\delta(t)$. For the general description of the system the inertial coordinate system I is introduced, which is fixed to the rigid track. Furthermore the accompanying coordinate systems C and K are introduced, which are fixed to the wheel center C. The purely translatory coordinate system C is suitable for the description of the contact, the rotating body-fixed coordinate system K is used for the formulation of the relative motion of the wheel structure.

For the computation of the dynamic displacements ${}^K\underline{\mathbf{u}}_g$, which are needed to calculate the end configuration $\delta(t)$ based on the reference configuration $\lambda(t)$, a FE model is used. The 2D model is meshed by a homogeneous discretization G. For this discretization 4-node elements with two displacements DOFs per node were chosen. A plane stress condition ("plane stress" setting of the element) is assumed. The density ρ_g shall be constant, furthermore a linear material law consisting of a constant Young modulus E_g and quasi-incompressibility $\nu_g = 0.49$ is assumed.

Initially no boundary conditions are defined. The geometrical modeling, the meshing and the assembly of the global system matrices is realized by the commercial finite-element program *ANSYS*. Eventually the stiffness matrix ${}^K\underline{\mathbf{K}}_g$ and the mass matrix ${}^K\underline{\mathbf{M}}_g$ are available in a body-fixed formulation. The required damping matrix is generated by a Rayleigh approach

$$
{}^K\underline{\mathbf{D}}_g = \alpha \, {}^K\underline{\mathbf{M}}_g + \beta \, {}^K\underline{\mathbf{K}}_g. \tag{6}
$$

The equations of motion of the wheel are set up within the inertial coordinate system I, because only there the absolute acceleration is available. However, at this point the presentation of the equations of motion within the inertial coordinate system I is skipped. Since the equations of the rolling contact model are formulated from a body-fixed point of view, the author rather emanates from equations of motion within body-fixed coordinate system K. The global $[g \times g]$ rotary matrix $\underline{\mathbf{A}}_g^{IK}$ and $\underline{\mathbf{A}}_g^{KI}$, respectively, required for the transformation between the coordinate systems, are gained from $[2 \times 2]$ elementary rotary matrices by diagonalization

$$\underline{\mathbf{A}}_g^{IK} = \mathrm{diag}(\underline{\mathbf{A}}^{IK}, \underline{\mathbf{A}}^{IK}, \ldots, \underline{\mathbf{A}}^{IK}), \tag{7}$$

further on $\underline{\mathbf{A}}_g^{KI} = (\underline{\mathbf{A}}_g^{IK})^T$ is imperative. The transformed equations of motion are

$$^K\underline{\mathbf{M}}_g \,{}^K_\delta \underline{\ddot{\mathbf{x}}}_g + {}^K\underline{\mathbf{D}}_g \,{}^K\underline{\dot{\mathbf{u}}}_g + {}^K\underline{\mathbf{K}}_g \,{}^K\underline{\mathbf{u}}_g = {}^K\underline{\mathbf{f}}_g^M, \tag{8}$$

where ${}^K\underline{\mathbf{f}}_g^M$ denotes the vector of contact loads acting on the mantle nodes of the wheel. Because of the plane modeling it has the dimension of a line load. The acceleration ${}^K_\delta \underline{\ddot{\mathbf{x}}}_g$ is applied and the right-hand side is adapted

$$^K\underline{\mathbf{M}}_g \,{}^K\underline{\ddot{\mathbf{u}}}_g + {}^K\underline{\mathbf{D}}_g \,{}^K\underline{\dot{\mathbf{u}}}_g + {}^K\underline{\mathbf{K}}_g \,{}^K\underline{\mathbf{u}}_g = {}^K\underline{\mathbf{f}}_g^M + {}^K\underline{\mathbf{f}}_g^G, \tag{9}$$

including the vector accounting for the inertia effects due to the absolute accelerated wheel

$$^K\underline{\mathbf{f}}_g^G = -{}^K\underline{\mathbf{M}}_g (\underline{\mathbf{E}}_g^C \underline{\mathbf{A}}^{KI}\,{}^I_\lambda \underline{\ddot{\mathbf{x}}}_C(t) + (\dot{\Omega}(t)\underline{\mathbf{E}}_g^t + \Omega^2(t)\underline{\mathbf{E}}_g^r)\,{}^K_\lambda \underline{\mathbf{x}}_g). \tag{10}$$

The system described in Equation (10) is still unconstrained. Since a given reference motion is impressed on the wheel, boundary conditions have to be determined. In reality the motion presetting is effected by a rigid hub, which is attached to the wheel structure. Due to this fix connection the motion of the hub nodes are consistent with the reference motion ${}^K_\delta \underline{\mathbf{x}}_n = {}^K_\lambda \underline{\mathbf{x}}_n$. This leads to

$$^K\underline{\mathbf{u}}_n = {}^K\underline{\dot{\mathbf{u}}}_n = {}^K\underline{\ddot{\mathbf{u}}}_n = \underline{\mathbf{0}}. \tag{11}$$

Thus, the corresponding n rows and columns are eliminated from the system matrices. The r displacements $\underline{\mathbf{u}}_r$ of the wheel nodes $\mathbf{R}$, consisting of structure nodes and mantle nodes, stay in the system of equations. Hereby the differential equation of the dimension r can be concluded

$$^K\underline{\mathbf{M}}_r \,{}^K\underline{\ddot{\mathbf{u}}}_r + {}^K\underline{\mathbf{D}}_r \,{}^K\underline{\dot{\mathbf{u}}}_r + {}^K\underline{\mathbf{K}}_r \,{}^K\underline{\mathbf{u}}_r = {}^K\underline{\mathbf{f}}_r^M + {}^K\underline{\mathbf{f}}_r^G. \tag{12}$$

The two vectors ${}^K\underline{\mathbf{f}}_r^M$ and ${}^K\underline{\mathbf{f}}_r^G$ on the right hand side result analogously from ${}^K\underline{\mathbf{f}}_g^M$ and ${}^K\underline{\mathbf{f}}_g^G$ by eliminating the n DOFs of the hub nodes $\mathbf{N}$.

3.1.2 Craig–Bampton Condensation

The large number of r DOFs in Equation (12) shall be replaced by a reasonable condensation method. A pure modal condensation would lead to a tremendous reduction of the number of DOFs and, by the diagonalization of the system matrices, to an additional decrease of the numerical complexity. But it would also implicate drawbacks. The time variant boundary

Experimental and Theoretical Investigations 233

condition within the sticking zone of the contact area would result in inaccurate solutions due to the stiff modal approximation. Furthermore, the local non-linear effects of the normal and the tangential contact would be smeared by the global ansatz functions. The modal superposition method for geometrically non-linear problems is explained in [21]. A pure static condensation would not be able to represent the dynamics of the system sufficiently due to its ansatz functions. Additionally the band structure of the system matrices is destroyed and thus the intended benefit of computational time would be balanced.

A mixed approach, which can be traced back to [4], combines the advantages of static and modal condensation. Local non-linearities in the contact area can be embedded, however, the number of DOFs stay in an acceptable range. Insight into the approach and also some application examples are provided by [6, 18]. Eventually, the differential equation Equation (12) can be transformed

$$^K\underline{\mathbf{M}}_h\,^K\underline{\ddot{\mathbf{u}}}_h + {}^K\underline{\mathbf{D}}_h\,^K\underline{\dot{\mathbf{u}}}_h + {}^K\underline{\mathbf{K}}_h\,^K\underline{\mathbf{u}}_h = {}^K\underline{\mathbf{f}}_h^M + {}^K\underline{\mathbf{B}}_h^C\underline{\mathbf{A}}^{KI}\,{}_\lambda^I\ddot{\mathbf{x}}_C(t) + {}^K\underline{\mathbf{B}}_h^t\,\dot{\Omega}(t) + {}^K\underline{\mathbf{B}}_h^r\,\Omega^2(t). \tag{13}$$

By using the Craig–Bampton condensation the number of DOFs can be reduced from r down to h. This corresponds to a reduction ratio of 80%.

3.1.3 Static Precalculation

During the simulation Equation (13) is solved by means of a time step integration. At each point in time t from the acceleration $^K\underline{\ddot{\mathbf{u}}}_h$ the velocity $^K\underline{\dot{\mathbf{u}}}_h$ and the displacement $^K\underline{\mathbf{u}}_h$ are calculated. The required initial conditions amount to

$$^K\underline{\dot{\mathbf{u}}}_h(t=0) = 0, \tag{14}$$

because angular velocity and hub velocity of the wheel are equal to zero $\Omega(t=0) = {}^Iv_C(t=0) = 0$ by definition. This does not apply for the initial node displacements $^K\underline{\mathbf{u}}_h(t=0) \neq 0$, since at the start of the simulation the wheel is already loaded by a normal force $^Kf_N^0$ und is thus predeformed. This initial normal force corresponds to the initial value of the setpoint setting

$$^Kf_N^0 = {}^C\bar{f}_N(t=0) \tag{15}$$

of the normal force feedback being an input variable of the global model. Since at the beginning of the simulation $t = 0$ the accompanying and the body-fixed coordinate system C and K, respectively, are orientated parallely, the usage of both coordinates in Equation (15) is allowable.

An overview on the operation of the static precalculation can be seen on Figure 8. The result of the precalculation serves as an initial condition for the integration of the displacement-vector $^K\underline{\mathbf{u}}_h(t=0) = {}^K\underline{\mathbf{u}}_h^0$. The normal force feedback needs an initial condition for the hub displacement

$$^Cw_C(t=0) = {}^Iw_C(t=0) = {}^Kw_C^0, \tag{16}$$

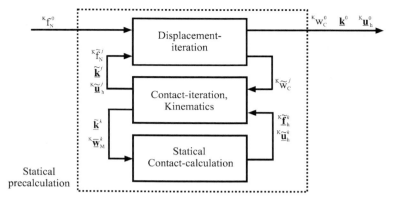

Fig. 8 Operation of the statical precalculation: The displacement-iteration calculates the hub displacement $^K w_C^0 (^K f_N^0)$ corresponding to the normal force presetting. In the contact-iteration for a given statical hub displacement $^K \tilde{w}_C$ the contacting nodes agreeable to the contact conditions are computed.

and the selection-algorithm for the contact-close nodes requires an initial contact configuration $\underline{k}^0$ as well. Since the inertial coordinate system I is orientated parallely to the accompanying coordinate system C for every point in time t, the equalization in Equation (16) is allowed. The contact indicator $\underline{k}^0$ contains the indices of the contacting nodes at the beginning of the simulation $t = 0$. In the field of numerical mechanics it is referred to as *Active Set*, cf. [30].

The actual contact calculation determines the statical displacement vector $^K \underline{\tilde{u}}_h$ and the corresponding contact load vector $^K \underline{\tilde{f}}_h$ for a given contact configuration, i.e. the static contact indicator $\underline{\tilde{k}}$ and the constrains $^K \underline{\bar{w}}_M$. The quantities of the statical contact calculation are furnished with a tilde in order to differentiate it from the dynamic displacements and the loads appearing later on in the chapter.

The statical precalculation is restricted to normal contact, no friction is assumed in tangential direction $\mu_0 = 0$. On the one hand the assumption of friction in the precalculation would be more realistic. But on the other hand the simulations in this chapter aim at the steady rolling process as well as at the unsteady changes between this steady processes. This phenomenons are, after a certain running-in-period (and also in terms of the linear modeling), independent from the initial condition of the system.

To find a distribution of the contacting nodes agreeable to the contact conditions, the statical precalculation is embedded in the contact-iteration. The results of the contact calculation are checked for plausibility and the contact indicator $\underline{\tilde{k}}^k$ as well as the constraint $^K \underline{\bar{w}}_M^k$ are adjusted. The superscripted index k marks the actual iteration step of the contact-iteration. The contact loads are summed up to the global normal force $^K \tilde{f}_N$. Thus a scalar result is

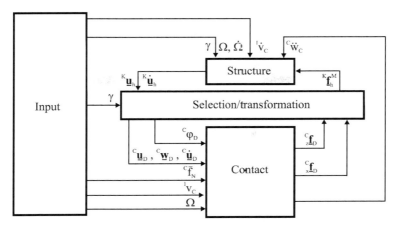

Fig. 9 Block diagram of the rolling contact simulation. The structure calculation operates with h DOFs of the condensated FE system, the contact calculation only uses d contact-close DOFs. The mapping of the DOFs is done in block *selection/transformation*.

generated for a given displacement of the wheel center ${}^K\tilde{f}_N({}^K\tilde{w}_C)$, suitable for the Newton's method in the displacement-iteration.

3.2 Simulation

After the considerations in Section 3.1 in this section the algorithms operating during the time step integration are explained. To clarify the relations between the individual modules and the flow of the computed quantities, an overview is given in the block diagram in Figure 9.

The block *structure* contains the condensated system of equations for the dynamic calculation, which was derived in Section 3.1.2. The system is excited by the contact loads ${}^K\underline{f}_h^M$ acting on the wheel mantle and the given absolute motion of the wheel, namely the angular velocity Ω und acceleration $\dot{\Omega}$, as well as the translational accelerations of the hub in horizontal and vertical direction ${}^I\dot{v}_C$ and ${}^I\ddot{w}_C$, respectively. This excitation results in dynamic displacements and displacement velocities of the wheel structure ${}^K\underline{u}_h$ and ${}^K\underline{\dot{u}}_h$, respectively.

In the block *selection/transformation* the contact-close nodes **D** are selected out of the set of mantle nodes **M** for every point in time t. This opens up the benefit, that the check of the contact conditions and the consequential calculation of the contact forces is only done for the nodes close to the contact area. This saves numerical resources and thus shortens the overall calculation time of the model.

The corresponding elements ${}^K\underline{u}_d$ are selected out of the displacement vector ${}^K\underline{u}_h$. Thereafter, the displacements are transformed into the accompanying coordinate system C, whereas the elements are divided into $^C x$-direction ($^C \underline{\mathbf{u}}_D$) and in $^C z$-direction ($^C \underline{\mathbf{w}}_D$). For the calculation of the tangential relative velocity within the contact area $^C \underline{\mathbf{v}}_D^{\text{rel}}$, analogously the vector of the displacement velocity $^K \underline{\dot{\mathbf{u}}}_h$ is decomposed, too. Furthermore, the reference position of the contact-close nodes $\mathbf{D}$, characterized by the angle $_\lambda^C \underline{\phi}_D$, is an output.

Within the block *contact* for every contact-close node D_i the conditions for normal contact are checked. In the case of normal contact a force is generated by means of the penalty method, so that a given constraint is almost kept. Moreover, for contacting nodes there are the two options of sticking and sliding in tangential direction. For sticking a constraint complying with global kinematics is calculated and analogously to the normal contact impressed on the mantle nodes by the penalty-method. For the calculation of this constraint the horizontal hub velocity $^I v_C$ and the angular velocity Ω of the wheel are required. In the case of sliding a friction force is computed, which acts on the system as an external force.

In the block *selection/transformation* the contact loads in tangential and normal direction $_x^C \underline{\mathbf{f}}_D$ and $_z^C \underline{\mathbf{f}}_D$ are transformed into the body-fixed coordinate system K. Afterwards they are assembled to the Craig–Bampton load vector $^K \underline{\mathbf{f}}_h^M$. A feedback loop within the *contact*-block provides the realization of the time dependent presetting of the normal force $^C \bar{f}_N$.

3.3 Identification of Parameters

In the previous part of the chapter the modeling is presented by means of general parameter values. Since the mechanical modeling in this section is validated by measurements, the physical based parameters of the model have to be identified or estimated, respectively.

3.3.1 Material

For the experiments in this chapter natural rubber (NR) was chosen as the material of the Grosch wheel, which has an almost linear-elastic behavior. This is confirmed by an uniaxial pulling test carried out by the *German Institute of Rubber Technology* (DIK), see Figure 10. In the experiment natural rubber only shows a very small Payne effect, and the Mullins effect is negligible. For the calculation of the material behavior therefore a linear isotropic material model, consisting of Young's modulus E_g and Poisson's number ν_g, is sufficient.

During the pulling test several strain cycles with the maximum strains $\epsilon_{\max} = 20, 50, 80, 100, 120, 140\%$ were run quasi-stationary. The undeformed cross-section of the specimen was used as the reference area for the stress computation independent from the strain level. The experiment shows a degressive stiffness behavior. Unlike a filled rubber material, no Mullins-effect can be identified during several runs of the same strain cycle. In fact the

Experimental and Theoretical Investigations

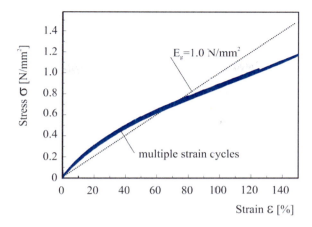

Fig. 10 Uniaxial pulling test for the determination of the Young modulus E_g of natural rubber (NR). Several strain cycles were done with different strain amplitudes.

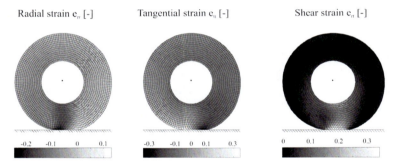

Fig. 11 Strain condition of the Grosch wheel at large slippage: Local strain in radial, tangential and shear direction for $S = 50\%$, $f_N = 40$ N, and $v_{\text{abs}} = 40$ mm/s.

measured cycles lie upon each other. Hysteresis loops can not be identified in the measurements as well.

In order to determine a resulting stiffness line in Figure 10 und thus a Young's modulus for the calculation, the strain condition of the wheel for a large slippage value $S = 50\%$ is regarded, see Figure 11. The figure separately shows the three strain components in radial, tangential and shear direction. The extreme values $\epsilon_{rr}^{\min} = -29\%$, $\epsilon_{tt}^{\max} = 39\%$ and $\epsilon_{rt}^{\max} = \epsilon_{tr}^{\max} = 38\%$ can be identified. The maximum occurring main strain amounts to $\epsilon_{I}^{\max} = 39\%$.

The quested line of best fit for a constant Young modulus in Figure 10 is determined based on the main strain value $\epsilon_{I}^{\max}$. However, since in further simulations with smaller slippage levels definitely higher strain values were identified (due to the sticking area in the contact zone), a Young modulus $E_g = 1.0$ N/mm² is defined, see Figure 10.

As in the pulling test in Figure 10 no hysteresis loops could be observed, according to [22] for the Rayleigh damping parameters rather small values are assumed $\alpha = 0$ und $\beta = 0.005$. The Poisson number is assumed to be $\nu_g = 0.49$, due to the incompressibility of rubber. The density of the applied material natural rubber amounts to $\rho_g = 1.2$ g/cm³. The inner diameter of the wheel model is chosen equal to the inner diameter of the Grosch wheel blank $r_N = 17.5$ mm. The Grosch wheel is mounted with two disks with an outer diameter 30 mm avoiding lateral extension, see. Figure 2. The complicated threedimensional stress condition within the clamped material is certainly best described by a modeling of the whole cross-section of the wheel.

3.3.2 Local Friction

The experiments shown in this chapter, see Section 2, were carried out on a glass surface. Due to the plane surface especially adhesion effects strongly come to the fore, thus the rubber friction is dominated by adhesion. For the determination of the local friction characteristic $\mu(v_{\rm rel})$ investigations with a clamped friction specimen on a rotating disc were carried out at the *Institute of Dynamics and Vibration Research*, see [27]. The maximum measured friction coefficient amounts to $\mu_{\max} \approx 1.2$, which is 50% of the maximum observed coefficient of friction $\mu^*_{\max}$ in the rolling contact experiments. The reason is the temperature dependency of the local friction coefficient $\mu(v_{\rm rel})$ of rubber. Unlike the "rotating" contact of the Grosch wheel, where the warmth is removed by the material flow, in the pure sliding contact of the applied test rig the rubber heats up strongly. In this case the heat generated by the friction process is only removed by thermal conduction. The resulting temperature of the specimen is much higher than the temperature of the rolling contact under comparable conditions of normal force and relative velocity. Since the local friction coefficient $\mu(v_{\rm rel})$ is dramatically lowered by increased temperatures, see for instance [14], steady measurements with a pure sliding contact are not suitable to identify a friction coefficient for rolling contact simulations.

For the curve of the local friction coefficient $\mu(v_{\rm rel})$ instead a simple Coulomb characteristic $\mu_0 = \mu^\infty$ is applied, see [26, 29]. Due to the discontinuity of the sliding friction value at $v_{\rm rel} = 0$ a non-smooth system arises, cf. [24]. This is resolved by a normalization, cf. [20]. The approach

$$\mu(v_{\rm rel}) = \mu^\infty \frac{2}{\pi} \arctan(s\, v_{\rm rel}) \qquad (17)$$

is chosen, where μ^∞ denotes the asymptotical sliding friction coefficient and s represents the slope parameter. In Figure 12 the normalized Coulomb characteristic is presented for various values of the slope parameter s, for the

Experimental and Theoretical Investigations 239

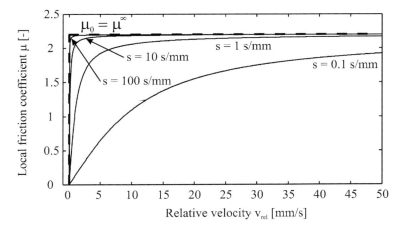

Fig. 12 Coulomb characteristic, consisting of stick value μ_0 and sliding value μ^∞ (dashed). The approximated, normalized curve is shown for various slope parameters $s = 0.1 \ldots 100$ s/mm.

simulation it was chosen to be $s = 1000$ s/mm. The identification of the static friction coefficient and the asymptotical sliding friction coefficient μ^∞ is done based on the steady measurements in Section 2.2. Corresponding to the maximum value at a normal force $F_N = 40$ N the friction coefficient is set to be $\mu_0 = \mu^\infty = \mu^*_{max} = 2.3$.

4 Results and Validations

During the steady rolling process the state variables show local constant characteristics, depending on the constant parameters slippage S and normal force $^C f_N$. Thus, the behavior of all contacting nodes on an arbitrary point in time can be suggested. In this way the position referenced characteristics of the individual nodal loads $^C_x f_{D,i}$ and $^C_z f_{D,i}$, respectively, the tangential node displacements $^C u_{D,i}$ or the relative velocity $^C v^{rel}_{D,i}$ deliver insights into the contact.

4.1 Steady Results

The position referenced characteristics of the tangential and the normal orientated node loads $^C_x f_{D,i}$ and $^C_z f_{D,i}$ strongly depend on the contact defining global parameters slippage S and normal force $^C f_N$. Therefore the dependency of the contact nodes on this global parameters are investigated in this section. Figure 13 shows the position referenced characteristics of the nodal loads under variation of the braking slippage $S = 10, 20, \ldots, 70\%$ at constant normal force $^C f_N = 40$ N.

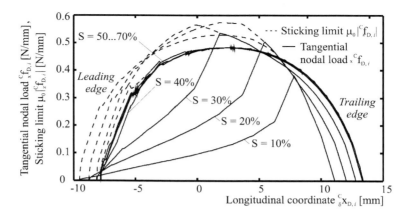

Fig. 13 Tangential nodal loads $^C_x f_{D,i}$ and the corresponding sticking limit $\mu_0|^C_z f_{D,i}|$ versus the longitudinal coordinate $^C_\delta x_{D,i}$ of the deformed configuration. The braking slippage amounts to $S = 10, 20, \ldots, 70\%$, the normal force is $^C f_N = 40$ N.

According to the coordinate systems introduced in Figure 7, the leading edge is labeled left in the diagram, the trailing edge is located on the right side. Instead of the normal nodal load $^C_z f_{D,i}$ the characteristic of the sticking limit $\mu_0|^C_z f_{D,i}|$ is shown in the diagram. In this way the diagram directly gives an information on the exhaustion of the friction connection and the exact point of separation.

Immediately after passing the leading edge, the contacting nodes initially stick to the track surface for $S < 50\%$. The shear caused by the global slippage leads to an increase of the tangential node load $^C_x f_{D,i}$ towards the trailing end of the contact area. For higher values of the slippage S the averaged slope of the tangential nodal load increases. For this reason the sticking limit $\mu_0|^C_z f_{D,i}|$ is reached earlier for higher slippage levels and thus the line of separation (start of sliding) moves towards the leading edge.

Uncoupled models like for instance the brush model, cf. [5], show a linear characteristic within the sticking zone. The unsteady rolling contact includes a coupled structure model and thus represents the influence of the neighboring nodes. This is proved by the strongly progressing slope of the tangential node load $^C_x f_{D,i}$. This coincides with both, continuous models, cf. [25], and tire measurements, cf. [8, 15].

If the tangential load of a contacting node $^C_x f_{D,i}$ reaches the sticking limit $\mu_0|^C_z f_{D,i}|$, the node separates from the track surface. According to *Coulomb's* friction law, the tangential load $^C_x f_{D,i}$ for sliding can be described by $\mu|^C_z f_{D,i}|$. Since the sticking and the sliding friction coefficients was set to be equal $\mu_0 = \mu$, see Section 3.3.2, the characteristic of the tangential load is identical to the sticking limit $\mu_0|^C_z f_{D,i}|$ within the sliding zone. For large slippage values $S > 50\%$ the sticking area in the contact zone disappears. The nodes

Experimental and Theoretical Investigations

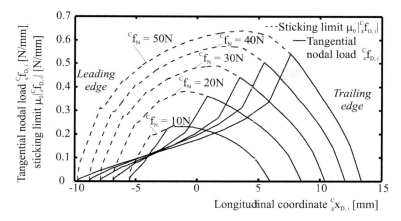

Fig. 14 Tangential node loads ${}^{C}_{x}f_{D,i}$ and the corresponding sticking limits $\mu_0|{}^{C}_{z}f_{D,i}|$ versus the longitudinal coordinate ${}^{C}_{\delta}x_{D,i}$ of the end configuration. The normal forces amount to ${}^{C}f_N = 10, 20, \ldots, 50$ N, the slippage level is $S = 20\%$.

incoming into the contact area are initially set to show sliding behavior. For small values of slippage $S < 50\%$ the incoming node D_i is decelerated by the sliding friction until the relative velocity ${}^{C}v^{\text{rel}}_{D,i} = 0$ is reached and the node sticks to the track surface. For higher slippage levels $S \geq 50\%$ the relative velocity of the incoming node D_i is too high, the deceleration in the contact area does not lead to the sticking transition anymore. Therefore no sticking zone exists for higher slippage levels: the entire contact area slides.

Figure 14 shows the dependency of the tangential load ${}^{C}_{x}f_{D,i}$ and the sticking limit $\mu_0|{}^{C}_{z}f_{D,i}|$ on the normal force ${}^{C}f_N$. The diagram presents the characteristics of the load ${}^{C}_{x}f_{D,i}({}^{C}_{\delta}x_{D,i})$ for the normal forces ${}^{C}f_N = 10$ N, 20 N, ..., 50 N, the slippage constantly amounts to $S = 20\%$. Within the sticking zone the slope of the tangential node load ${}^{C}_{x}f_{D,i}({}^{C}_{\delta}x_{D,i})$ decreases for increasing normal forces ${}^{C}f_N$. The reason for this phenomenon is the larger number of contacting nodes for higher normal forces ${}^{C}f_N$. Although the transferred tangential force ${}^{C}f_T$ increases, the number of contacting nodes increases even more for higher normal forces ${}^{C}f_N$. If a displacement is impressed on a contact node within the sticking zone, for higher normal forces ${}^{C}f_N$ a smaller portion of the tangential force ${}^{C}f_T$ falls upon an individual node. This leads to decreasing slopes of the characteristic ${}^{C}_{x}f_{D,i}({}^{C}_{\delta}x_{D,i})$ for increasing normal forces ${}^{C}f_N$. The higher slope values of the tangential node load ${}^{C}_{x}f_{D,i}({}^{C}_{\delta}x_{D,i})$ result in the fact, that the friction connection on the same slippage level $S = 20\%$ is already exhausted for small normal forces ${}^{C}f_N$.

For a slippage $S = 20\%$, a normal force ${}^{C}f_N = 10$ N and a maximum friction coefficient $\mu_{\max} = 2.3$ already a force ratio $\mu^* \approx 2.1$ is computed. A higher normal force ${}^{C}f_N = 50$ N results in $\mu^* \approx 1.2$ and thus the friction connection is considerably less exhausted. Along the lines of Figure 13 the

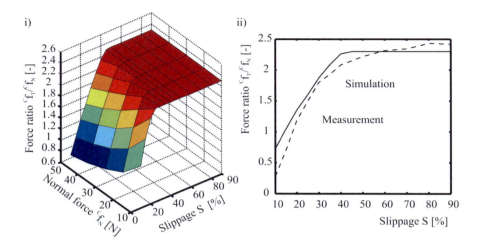

Fig. 15 (i) Simulated characteristic map of the force relation $^Cf_T/^Cf_N$ depending on normal force ^Cf_N and slippage S. (ii) Section through the map at $^Cf_N = 40$ N, comparison between measurements and simulations.

characteristic $^C_xf_{D,i}(^C_\delta x_{D,i})$ is shifted towards the positive direction of the abscissa. the reason is that for increasing normal forces ^Cf_N still higher values of the tangential force ^Cf_N are transferred.

4.1.1 Steady Results: Comparison with Measurements

After the local insights into the rolling contact, a comparison between the steady measurements in Section 2.2 and the simulation results are carried out by means of global quantities. Figure 15(i) shows a characteristic map concluding the results of the steady simulations. The map presents the dependency of the force relation $^Cf_T/^Cf_N$ on normal force ^Cf_N and slippage S. Analogous to the steady measurements on Figure 3, the simulations were carried out in the range of $^Cf_N = 10, 20, \ldots, 50$ N and $S = 10, 20, \ldots, 90\%$.

The force relation $^Cf_T/^Cf_N$ initially increases and then reaches the saturation value $^Cf_T/^Cf_N = 2.3$. This value corresponds to the sliding friction coefficient μ^∞ defined in Section 3.3.2. For large values of slippage S the entire contact area slides, so that consequently $^Cf_T/^Cf_N = \mu^\infty$ is effective. For smaller values of slippage S parts of the contact area stick to the track surface. This nodal loads act like constraining loads, which can be at most equal to the sticking limit $\mu_0 |^C_z f_{D,i}|$. Since sticking and sliding value of *Coulomb's* law have been equalized $\mu_0 = \mu$, the potential maximum of the force relation amounts to $(^Cf_T/^Cf_N)^{\max} = 2.3$ as well. Hence a saturation behavior results, no clear maximum can be identified.

Furthermore, the force relation $^Cf_T/^Cf_N$ shows a falling dependency on the normal force ^Cf_N. This effect is based mainly on the decreasing shear stiffness of the wheel for increasing normal forces, see Figure 14. The dependencies of the force relation $^Cf_T/^Cf_N$ on the normal force ^Cf_N and the slippage S observed in the simulations also appear in the measurements of the natural rubber material shown in Figure 3. However, the falling dependency on the normal force ^Cf_N is less significant in the measurements.

Figure 15(ii) shows the comparison between measurements and calculation for a section of the characteristic map at $^Cf_N = 40$ N. The measured curve starts with $F_T/F_N \approx 0.3$ for $S = 10\%$, the simulation results in a higher force relation $^Cf_T/^Cf_N \approx 0.7$ at the same slippage level. For increased slippage S both characteristics show a saturation behavior, whereas the measured curve appears to be rounder. In contrast, the graph of the simulated force relation increases almost linearly within the interval $S \leq 40\%$, for $S > 40\%$ the simulation results in nearly constant values $^Cf_T/^Cf_N \approx 2.3$. This "hard" characteristic in the simulation is due to *Coulomb's* law, which represents only one constant value μ for sliding friction. With a rubber-specific friction law, for the local friction $\mu(v_\text{rel})$, the difference between measurement and simulation could be further minimized.

4.2 Unsteady Results

For the steady simulations, the comparison with measurements showed a good accordance. The presented contact model is suitable not only for simulations with steady rolling, but also for calculations with an transient excitation of the friction contact. Therefore, the system behavior in consequence of slippage steps is investigated in this section. The results will eventually be compared to the measured quantities, which have already been shown in Section 2.3.2.

4.2.1 Time Series

The unsteady measurements in Section 2.3.2 show a dependency of the tangential force $^Cf_T(t)$ on the start and the target level of the slippage step. Figure 16 shows the corresponding simulated time series.

The upper diagrams show slippage steps from $S_\text{start} = 10\%$ up to a higher level (left) and the corresponding return steps down to $S_\text{target} = 10\%$. The lower diagrams analogously show steps down from $S_\text{start} = 70\%$ and back to $S_\text{target} = 70\%$. In all cases the normal force amounts to $^Cf_N = 40$ N, at a hub velocity $^Cv_C = 40$ mm/s of the wheel.

The maximal tangential force $^Cf_T \approx 92$ N is achieved for slippages $S \geq 50\%$. This is plausible, because for *Coulomb's* law a sliding friction coefficient $\mu = 2.3$ was assumed. Regarding the time series in Figure 16, some tendencies, which have already been observed in the measurements, can be confirmed. For slippage steps with the start level $S_\text{start} = 10\%$ (above, left) the time constant of the tangential force $^Cf_T(t)$ decrease for increasing target levels S_target. Regarding steps from $S_\text{start} = 70\%$ (below, left) down to a lower level, the same tendency can be identified.

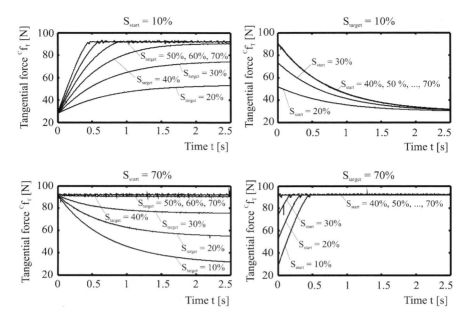

Fig. 16 Simulated time series of the tangential force $^C f_T(t)$ after a slippage step at constant normal force $^C f_N = 40$ N and hub velocity $^C v_C = 40$ mm/s. The diagrams show slippage steps with the start- and target levels $S_\text{start} = 10\%$ (above, left) and $S_\text{target} = 10\%$ (above, right) and $S_\text{start} = 70\%$ (below, left) and $S_\text{target} = 70\%$ (below, right).

Considering the steps back to the base level $S_\text{target} = 10\%$ (above, right) generally higher time constant for smaller start levels S_start of the slippage step can be observed. In this diagram, for start levels $S_\text{start} \geq 40\%$ no change for the time constants can be identified, since due to *Coulomb's* law this slippage levels all result in the same tangential force $^C f_T$. Also for slippage steps up to $S_\text{target} = 70\%$ (below, right) the tendency is identical, however, the absolute values of the time constant are considerable lower than for $S_\text{target} = 10\%$.

4.2.2 Unsteady Results: Comparison with Measurements

Analogous to the procedure for the characterization of the measured tangential force time series $F_T(t)$, the time constant τ is also determined for the simulated curves $^C f_T(t)$, cf. Section 2.3.2. Like in the measurements, a clearly arranged overview of the results is achieved by presenting the time constant τ versus the start and the target level S_start and S_target, respectively. Figure 17 shows two diagrams, where the slippage levels S_start and S_target are shown as parameter and as independent variable on the abscissa, respectively. The same parameter range was chosen as in the experimental figures, see Figures 5 and 6.

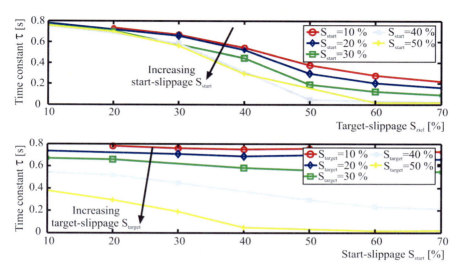

Fig. 17 Time constant τ for simulated tangential force time series $^C f_T(t)$ versus the target and start level of the slippage step S_{target} (above) and S_{start} (below), respectively

The upper diagram shows the time constant τ plotted versus the target level S_{target} of the slippage step. Like in the measurements, the time constant τ generally decrease for increasing start levels S_{start}. The highest value for the time constant $\tau \approx 0.8$ s is achieved for a slippage step from $S_{\text{start}} = 20\%$ down to $S_{\text{target}} = 10\%$. In the range $10\% \leq S_{\text{target}} \leq 50\%$ the set of curves shows a parabolic behavior. Whereas for $S_{\text{target}} = 10\%$ the individual graphs are quite similar, for larger target values the curves diverge. For $S_{\text{target}} > 50\%$ the graphs slightly kink and decrease almost linearly with a small slope. This qualitative developing was not found in the measurements, the measured characteristics decrease over the entire investigated range nearly linearly. The kink in the simulated curves is due to *Coulomb's* "hard" friction characteristic, which gives, compared to real rubber friction behavior, the maximum value of the tangential force already for very small relative velocities.

The lower diagram of Figure 17 shows the relation between the time constant τ of the tangential force time series and the start level of the slippage step S_{start}. The falling dependency of the time constant τ on the start level S_{start} is considerably lower than the influence of the target level S_{target}. This tendency was already observed in the experimental investigations. The curve starts for a target level $S_{\text{target}} = 10\%$ with the time constant $\tau_{20,\,10} \approx 0.78$ s and only falls down slightly to $\tau_{70,\,10} \approx 0.73$ s. For high values of the target slippage S_{target} the decay increases: the characteristic of $S_{\text{target}} = 40\%$ falls from $\tau_{10,\,40} \approx 0.55$ s down to $\tau_{70,\,40} \approx 0.22$ s.

The characteristic $\tau(S_{\text{start}})$ for a target level $S_{\text{target}} = 50\%$ shows a kink at $S_{\text{start}} = 40\%$. Like in the upper diagram, which also shows a kink at

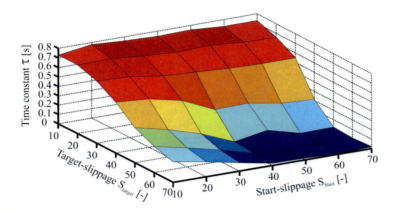

Fig. 18 Time constant τ depending on the start and the target level S_{start} and S_{target}.

$S_{\text{target}} = 50\%$, the kink is because of the for rubber unrealistic friction behavior of *Coulomb's* law.

Compared to the results of the unsteady measurements shown in Figures 5 and 6 the rolling contact model represents the tendencies of the real rolling friction very well. Hence, particularly the rolling kinematics represented by the model play a role for the observed time delay between step excitation and frictional response. Some rubber-specific characteristics, like for instance the temperature dependency of the elasticity and the local friction, have not been regarded, see [7]. However, the simulated and the measured characteristics of the time constant τ differ by a factor 2. Hence, the simulation shows a stiffer behavior than the real rolling contact.

The simulation is based on the assumption that the test rig moves along with a constant velocity. In reality the vehicle velocity is controlled by the actuators of the test rig. The slippage step disturbs the velocity of the test rig, which again causes a reaction of the rolling contact. The overall dynamics of the velocity controller approximately lie in the range of the observed friction time constants. Consequently, the longitudinal dynamics of the test rig lead to a delay of the Grosch wheel's behavior, which is not represented in the simulations.

A certain uncertainty of the rolling contact model is given by the 2D modeling of the Grosch wheel structure. The realistic 3D stress condition particularly between contact and hub is approximated by a 2D 4-node element with a *plane stress* approach. Although this approximation is the best possible, it results in a stiffer system behavior, compared to a 3D modeling. Also the already mentioned stiff friction characteristic of the local friction coefficient $\mu(v_{\text{rel}})$ plays a role for the lowering of the time constants in the simulation.

Figure 18 gives an overview on the dependency of the time constant τ of the tangential force time series $^C f_T(t)$ on the start and the target level S_{start}

and S_{target}, respectively. The falling dependency of the time constant τ on both, the start and the target level can be clearly seen. In the range of high slippages the time constant gets very small, because there is no difference in the resulting tangential force anymore due to the applied friction law.

5 Conclusions

In many technical applications rolling contacts of deformable bodies can be found. For rapid changes of slippage or normal force the contact conditions cannot be described by steady models anymore. For this purpose a model for unsteady rolling contact is developed, representing the viscoelastic material behavior of the contacting body. The model is realized by means of a small solid rubber wheel ("Grosch wheel") and a rigid glass trackway. The relevant algorithms can be transferred to other technical fields, as for instance printing machines or band conveyors.

In this article steady measurements are initially carried out with of a specially constructed moving test rig. Thereby wheels consisting of various filled and unfilled materials are applied. The material natural rubber shows the best properties for the unsteady measurements. For the characterization of the dynamical behavior after slippage steps an exponential saturation behavior is fitted to the measurements. The time constant of the saturation behavior acts as a quantity for the dynamics of the rolling contact. Eventually the dependency of this time constant on the start and the target level of the slippage step at a constant normal force is presented. For increasing start levels of the slippage step the time constant decreases. For increasing values of the target level the time constant decreases as well. Here the dependency of the time constant on the target level of the slippage step is the considerably stronger influencing factor on the friction dynamics. The highest values of the time constant consequently appear for small start and target levels, namely $\tau \approx 1.5$ s. The minimal identified time constant amounts to $\tau \approx 0.1$ s.

The phenomenons observed in the steady and unsteady measurements are represented by the 2D rolling contact model. The representation of the wheel structure is based on a Finite-Element description. The large number of DOFs is strongly reduced by a Craig–Bampton transformation. The master DOFs of this transformation are the displacements of the wheel mantle, all other DOFs are concentrated by means of a modal approach. If a node gets into the contact area, a normal load is impressed with the help of the penalty-method minimizing the penetration of the node into the trackway. In tangential direction, the model differentiates between sticking and sliding friction. In case of sticking the displacement of the node, depending on the global quantities hub velocity and angular velocity of the wheel, is impressed analogously to the normal contact applying the penalty method. If the node slides, a friction force depending on the local relative velocity and normal load is calculated.

The comparison of the steady friction characteristic works well, whereas the small deviations could be further minimized with a rubber-specific characteristic of the local friction coefficient. Like in the measurements, the system behavior is assessed by means of time constants, characterizing the contact dynamics immediately after a slippage step. The dependencies of this time constant on the start and target level of the slippage step observed in the measurements also appear in the results of the rolling contact model. However, the results of measurement and calculation differ approximately by the factor 2. The reason is that some effects like temperature effects resulting in an additional time constant are not considered in this model. The characteristics of the nodal loads provided by the calculation give insights into the local rolling contact behavior. The presented results are qualitatively consistent with tire-specific characteristics known in the literature.

References

1. Blume, H., Gutzeit, F.: Serviceroboter ermittelt Reibwerte von Gummi. Technologie-Informationen niedersächsischer Hochschulen 2004(3), 6 (2004)
2. Blume, H., Heimann, B., Lindner, M.: Design of an outdoor mobile platform for friction measurement on street surfaces. In: Proceedings of the International Colloquium on Autonomous and Mobile Systems, Magdeburg, pp. 65–68 (2002)
3. Blume, H., Heimann, B., Lindner, M., Volk, H.: Friction measurements on road surfaces. Kautschuk Gummi Kunststoffe 56(12), 677–681 (2003)
4. Craig, R.R., Bampton, M.C.C.: Coupling of substructures for dynamic analysis. AIAA Journal 6(7), 1313–1319 (1968)
5. Fromm, H.: Berechnung des Schlupfes beim Rollen deformierbarer Scheiben. Zeitung für angewandte Mathematik und Mechanik 7(1), 27–58 (1927)
6. Gasch, R., Knothe, K.: Strukturdynamik. Band 2: Kontinua und ihre Diskretisierung. Springer, Berlin (1989)
7. Gäbel, G., Kröger, M.: Reasons, models and experiments for unsteady friction of vehicle tires. In: VDI Berichte, vol. 2014, pp. 245–259 (2007)
8. Gerresheim, M., Hussmann, A.W.: Kräfte und Bewegungen in der Aufstandsfläche geradeausrollender Reifen. Automobiltechnische Zeitschrift 77(6), 165–169 (1975)
9. Grosch, K.A.: Laborbestimmung der Abrieb- und Rutschfestigkeit von Laufflächenmischungen – Teil I: Rutschfestigkeit. Kautschuk Gummi Kunststoffe 49(6), 432–441 (1996)
10. Gutzeit, F., Kröger, M., Lindner, M., Popp, K.: Experimental investigations on the dynamical friction behaviour of rubber. In: Proceedings of 6th Rubber Fall Colloquium, Hannover, pp. 523–532 (2004)
11. Gutzeit, F., Sextro, W., Kröger, M.: Unsteady rolling contact of rubber wheels. In: Analysis and Simulation of Contact Problems, pp. 261–270. Springer, Berlin (2006)
12. Gutzeit, F., Wangenheim, M., Kröger, M.: An experimentally validated model for unsteady rolling. In: III European Conference on Computational Mechanics, p. 320. Springer, Dordrecht (2006)

13. v. Huffel, S., Lemmerling, P.: Total Least Squares and Errors-in-Variables Modeling. Analysis, Algorithms and Applications. Springer, Berlin (2001)
14. Kröger, M., Wangenheim, M., Moldenhauer, P.: Temperatureffekte auf das lokale Reibverhalten von Elastomeren im Gleit- und Rollkontakt. In: 10. Int. Tagung Reifen-Fahrwerk-Fahrbahn, VDI Berichte, pp. 271–290. VDI-Verlag, Düsseldorf (2005)
15. Kummer, H.W., Meyer, W.E.: Verbesserter Kraftschluß zwischen Reifen und Fahrbahn – Ergebnisse einer neuen Reibungstheorie. Automobiltechnische Zeitschrift 69(8), 245–251 (1967)
16. Lawson, C.L., Hanson, R.J.: Solving Least Square Problems. Society for Industrial and Applied Mathematics, Philadelphia (1995)
17. Lindner, M.: Experimentelle und theoretische Untersuchungen zur Gummireibung an Profilklötzen und Dichtungen. Fortschritt-Berichte 11(331) (2006)
18. Link, M.: Finite Elemente in der Statik und Dynamik. Band 2: Kontinua und ihre Diskretisierung. 3rd (ed.), B.G. Teubner Verlag, Wiesbaden (2002)
19. Meyer, W.E., Kummer, H.W.: Die Kraftübertragung zwischen Reifen und Fahrbahn. Automobiltechnische Zeitschrift 66(9), 245–250 (1964)
20. Mostaghel, N., Davis, T.: Representations of Coulomb friction for dynamic analysis. Earthquake Engineering and Structural Dynamics 26(5), 541–548 (1997)
21. Nickell, R.E.: Nonlinear dynamics by mode superposition. Computer Methods in Applied Mechanics and Engineering 7, 107–129 (1976)
22. Ottl, D.: Schwingungen mechanischer Systeme mit Strukturdämpfung, Forschungsbericht, No. 603. VDI Verlag, Düsseldorf (1981)
23. Pacejka, H.B.: Modelling of the Pneumatic Tire and its Impact on Vehicle Dynamic Behaviour, Carl-Cranz-Gesellschaft e.V, Oberpfaffenhofen (1992)
24. Popp, K.: Non-smooth mechanical systems – An overview. Forschung im Ingenieurwesen 64(9), 223–230 (1998)
25. Popp, K., Schiehlen, W.: Fahrzeugdynamik. Eine Einführung in die Dynamik des Systems Fahrzeug-Fahrweg. B.G. Teubner, Stuttgart (1993)
26. Sasada, T., Nakabayashi, H.: Does Hertzian contact area act as an effective zone generating the friction resistance? IEICE Transactions on Electronics E81-C(3), 326–329 (1998)
27. Sextro, W., Moldenhauer, P., Wangenheim, M., Lindner, M., Kröger, M.: Contact behaviour of a sliding rubber element. In: Wriggers, P., Nackenhorst, U. (eds.) Analysis and Simulation of Contact Problems. LNACM, vol. 27, pp. 243–252. Springer, Heidelberg (2006)
28. Sjahdanulirwan, M.: An analytical model for the prediction of tyre-road friction under braking and cornering. International Journal of Vehicle Design 14(1), 78–99 (1993)
29. Stein, G.: Reibungsverhalten von Elastomerwerkstoffen und -bauteilen. Technische Problemlösungen mit Elastomeren, Reihe VDI-K, pp. 1–23. VDI-Verlag, Düsseldorf (1992)
30. Wriggers, P.: Computational Contact Mechanic. John Wiley and Sons, Hoboken (2002)

Author Index

Besdo, D.   95
Bouzid, Noamen   201
Busse, L.   1

Gutzeit, F.   221
Gvozdovskaya, N.   95

Heimann, Bodo   201

Klüppel, M.   1, 27
Kröger, Matthias   165, 221

Le Gal, A.   1
Lorenz, H.   27

Meier, J.   27
Moldenhauer, Patrick   165

Nackenhorst, U.   123

Oehmen, K.H.   95

Reinelt, Jana   53

Suwannachit, A.   123

Wriggers, Peter   53

Ziefle, M.   123